Communications
in Computer and Information Science 636

Commenced Publication in 2007
Founding and Former Series Editors:
Alfredo Cuzzocrea, Dominik Ślęzak, and Xiaokang Yang

More information about this series at http://www.springer.com/series/7899

Hongxiu Li · Pirkko Nykänen
Reima Suomi · Nilmini Wickramasinghe
Gunilla Widén · Ming Zhan (Eds.)

Building Sustainable Health Ecosystems

6th International Conference
on Well-Being in the Information Society, WIS 2016
Tampere, Finland, September 16–18, 2016
Proceedings

Springer

Editors

Hongxiu Li
University of Turku
Turku
Finland

Nilmini Wickramasinghe
Deakin University
Melbourne
Australia

Pirkko Nykänen
University of Tampere
Tampere
Finland

Gunilla Widén
Åbo Academi University
Turku
Finland

Reima Suomi
University of Turku
Turku
Finland

Ming Zhan
Åbo Academi University
Turku
Finland

ISSN 1865-0929   ISSN 1865-0937   (electronic)
Communications in Computer and Information Science
ISBN 978-3-319-44671-4   ISBN 978-3-319-44672-1   (eBook)
DOI 10.1007/978-3-319-44672-1

Library of Congress Control Number: 2016948279

Printed on acid-free paper

This Springer imprint is published by Springer Nature
The registered company is Springer International Publishing AG Switzerland

Preface

The Well-being in the Information Society (WIS) conference series celebrated its sixth biennial event this year. Started in 2006, the conference seems to be more at the heart of our societal and scientific needs than ever before. Much has been learnt during the first 10 years of the conference series and soon we will enter the teenage years, where new avenues and visions will surely emerge. Issues such as the global refugee crisis, rapid urbanization, circular economy, increasing digitalization, all-encompassing globalization, and a continually evolving understanding and knowledge of health and well-being are all at the very core of the long-term WIS agenda.

A key idea of the WIS conference series has been to take a multidisciplinary approach to well-being and rural and information society problems, as is witnessed by this year's program. A second key idea has been to encourage interaction between researchers and people doing daily practical work in the areas mentioned. The core of the conference has been scientific discussion documented in peer-reviewed scientific papers, but there has always been room for other contributions, too. This year we were very happy to be able to anchor our conference place and time to the WHO World Conference Safety 2016, whilst remaining an independent conference taking place beside that mega-event.

Well-being in the Information Society - Building Sustainable Health Ecosystems (WIS 2016) focused on innovations and fresh ideas at the cross-section of urban living, the information society, and health as understood in a wide sense.

We received high-quality contributions, based both on scientific research and practical experience. These papers clearly demonstrate the multidisciplinary nature of the conference. The accepted papers were organized within the following seven broad topics: 1. Macro level considerations of e-health and welfare, 2. Welfare issues of children, adolescents, young elderly, and seniors, 3. Analytics issues of e-health and welfare, 4. National/regional initiatives in e-health and welfare, and 5. Specific topics of e-health. The papers grouped within these topics span qualitative and quantitative analysis, empirical surveys, case studies, as well as conceptual work.

We are proud to have had very qualified, international keynote speakers at our conference: Dr. Ramesh Krishnamurthy, WHO; Prof. Kshanika Hirimburegama, University of Colombo, Sri Lanka; Dr. J. Marc Overhage, Chief Medical Informatics Officer for Cerner; Dr. Olli Kangas, The Social Insurance Institution of Finland (Kela); Eva Jakobson Vaagland, Norwegian Safety Forum; Prof. Peeter Ross, Tallinn University of Technology; and Prof. Nilmini Wickramasinghe, Epworth Health Care and Deakin University, Australia. These keynote speakers presented cutting-edge information and innovation from the field.

Together with the keynotes, the accepted papers offered to all participants new perspectives and insights into sustainable health ecosystems, their drivers and barriers, theoretical foundations and methodologies, users' perspectives and user experiences, as well as national health IT ecosystems and their development. Additionally, we held panels and workshops on relevant topics such as ecosystems in health information systems procurement, technology as a resource in elderly care, trust in digitalization, and big cities meet big data. All these contributions make the WIS 2016 conference a perfect forum for the exchange of information and ideas.

WIS 2016 was organized by Turku School of Economics and Faculty of Medicine, University of Turku and Baltic Region Healthy Cities Association – WHO Collaborating Centre for Healthy Cities and Urban Health in the Baltic Region in co-operation with Åbo Akademi University; Turku University of Applied Sciences, the Finnish National Institute of Health and Welfare, Kela, the University of Tampere; and the University of Eastern Finland.

We wish to thank all the authors of the proceedings for sharing their research and innovation, as well as all the reviewers who have contributed to the quality of these proceedings and the whole conference. We would not have been able to run this conference without the support of our organizations, and the proceedings would look different without the kind support of Springer; thank you to all these parties. WIS 2016 gained financial support from the Foundation for Economic Education, the Federation of Finnish Learned Societies, and Åbo Akademi Foundation. We thank all these parties for their support.

We hope you all enjoyed a challenging, innovative, and fruitful conference with keynotes, scientific paper sessions, panels and the social program as well as good moments and new insights later when reading these proceedings.

June 2016

Hongxiu Li
Pirkko Nykänen
Reima Suomi
Nilmini Wickramasinghe
Gunilla Widén
Ming Zhan

Organization

WIS 2016 Organizing Committee

Suomi, Reima (Conference Chair) — University of Turku, Finland

Li, Hongxiu (Organizing Committee Chair) — University of Turku, Finland

Nykänen, Pirkko (Program Co-chair) — University of Tampere, Finland

Wickramasingne, Nilmini (Program Co-chair) — Deakin University, Australia

Karlsson, Marlene — Turku University of Applied Science, Finland

Kärkkäinen, Jukka — The National Institute for Health and Welfare, Finland

Machiewicz, Karolina — Baltic Region Healthy Cities Association, Finland

Putkinen, Marjut — Turku University of Applied Sciences, Finland

Reiman, Johanna — Baltic Region Healthy Cities Association, Finland

Salanterä, Sanna — University of Turku, Finland

Somerkoski, Brita — University of Turku, the National Institute for Health and Welfare, Finland

Widén, Gunilla — Åbo Akademi University, Finland

Zhan, Ming — Åbo Akademi University, Finland

WIS 2016 Program Committee

Amdam, Roar — Volda University College, Norway

Ammenwerth, Elske — University for Health Sciences, Austria

Bergum, Svein — Lillhammer University College, Norway

Borlund, Pia — University of Copenhagen, Denmark

Brender, Jytte — Aalborg University, Denmark

Cabral, Regis — FEPRO - Funding for European Projects, Sweden

Carmichael, Laurence — The University of the West of England, UK

Cellary, Wojceich — Poznan University of Economics, Poland

Christa, Womser-Hacker — University of Hildesheim, Germany

D'Cruz, Martin — Indiana University, USA

Deursen, Nicole — Edinburgh Napier University, UK

Georgiou, Andrew — Macquarie University, Australia

Harnesk, Dan — Luleå University of Technology, Sweden

Hansen, Preben — Stockholm University, Sweden

Hori, Mayumi — Hakuoh University, Japan

Ison, Erica — Right Care, UK

Jackson, Paul	Oxford Brookes University, UK
Järveläinen, Jonna	University of Turku, Finland
Kalvenas, Alvydas	Lithuanian Sports University, Lithuanian
Karlsson, Marlene	Turku University of Applied Sciences, Finland
Kini, Ranjan	Indiana University Northwest, USA
Klein, Stefan	University of Muenster, Germany
Kokol, Peter	University of Maribor, Slovenia
Koskinen, Jani	University of Turku, Finland
Krebs, Irene	BTU, Cottbus, Denmark
Kurland, Lisa	Karolinska Institutet, Sweden
Lamersdorf, Winfried	University of Hamburg, Germany
Lawrence, Roderick	University of Geneva, Switzerland
Li, Hongxiu	University of Turku, Finland
Liu, Yong	Aalto University School of Business, Finland
Maciaszek, Leszek	Wrocław University of Economics, Poland
Mäntymäki, Matti	University of Turku, Finland
Mettler, Tobias	University of St. Gallen, Switzerland
Moen, Anne	University of Oslo, Norway
Moring, Camilla	University of Copenhagen, Denmark
Müller, Oliver	University of Liechtenstein, Liechtenstein
Nykänen, Pirkko	University of Tampere, Finland
Paavilainen, Eija	University of Tampere, Finland
Paget, Dineke	EUPHA, Netherlands
Palsdottir, Agusta	University of Iceland, Iceland
Papacharalambous, Lefkada	Technological Educational Institution of Sterea Ellada, Greece
Päivärinta, Tero	Luleå University of Technology, Sweden
Rapp, Birger	IMIT, Sweden
Rasmussen, Niels Kristian	Ostfold County Council, Denmark
Rigby, Michael	Keele University, UK
Rimantas, Butleris	Kaunas University of Technology, Lithuania
Ryjov, Alexander	Moscow State University, Russia
Salmela, Hannu	University of Turku, Finland
Saru, Essi	University of Turku, Finland
Schellhammer, Stefan	University of Münster, Germany
Scott, Philip	University of Portsmouth, UK
Sheerin, Fintan	University of Dublin, Ireland
Singh, Rajesh	Emporia State University, USA
Sjöblom, Olli	University of Turku, Finland
Suomi, Reima Vesa	University of Tuku, Finland
Thomas, Mandl	University of Hildesheim, Germany
Vähämäki, Maija	University of Turku, Finland
Villerusa, Anita	Riga Stradins University, Latvia
Vom Brocke, Jan	University of Liechtenstein, Liechtenstein
Webster, Premila	University of Oxford, UK
Wells, George	Rhodes University, South Africa

Wickramasinghe, Nilmini	Epworth HealthCare and Deakin University, Australia
Widén, Gunilla	Åbo Akademi University, Finland
Wrycka, Stanislaw	University of Gdansk, Poland
Österle, Hubert	University of St. Gallen, Switzerland

Contents

Macro Level Considerations of e-Health and Welfare

Bridging the Gap - Health, Technology and Intermediaries

Janne Lahtiranta[✉]

Department of Information Technology, University of Turku, Turku, Finland
janne.lahtiranta@it.utu.fi

Abstract. The demands for health care services are rising steadily. To meet these demands, the prevailing health care paradigm has been put under scrutiny. We have already passed the verge of a paradigm shift where patients are regarded cooperative partners who play an important role in their own care. In the core of this ongoing shift is technology that enables a new kind of service provisioning and digitalization of services. This has created a dilemma; how the patients who are not able (or willing) to use these new services can be reached? To answer this question, a fresh outlook on the roles related to health service provisioning is needed. This examination calls for identification of health technology intermediaries, mediators, who stand at the crossroad of health, care and technology. This exploratory work will look into the intermediary roles, outlining a set of skills and capabilities needed from a mediator.

Keywords: Health care information systems · Health service provisioning · Intermediaries

1 Introduction

It is a commonly accepted view that the field of health care is changing in fundamental ways. This change is put into motion by economic and societal drivers, such as ageing societies [1], which are affecting all levels of the field, from funding to practicalities of care. Probably one of the most visible trends associated with the ongoing change is related to re-delegation of care. Instead of placing responsibilities of care into the hands of a health service provider, they are gradually shifted towards the patient (or one's relative). This re-delegation is particularly evident if we look into the patient-physician relationship, which has changed fundamentally during the last two decades. While the traditional (or paternalistic) relationship was one of the most prevalent models until the turn of the millennium, other models have gained a more foothold since then. These models, which are in general more equal in terms of decision making, and more consumer-oriented reflecting the spirit of our time, include the partnership model and the self-governing model. In the first one the actors (patient and physician) are considered equal and the patient has more responsibilities than in a traditional model [2]. In the second one, the patient is even in a more prominent position as the relationship is akin to the one between a buyer and a seller [2]. Key enablers in this change are electronic health services that alter geographical and spatial dimensions of care. In practice, these

© Springer International Publishing Switzerland 2016
H. Li et al. (Eds.): WIS 2016, CCIS 636, pp. 3–14, 2016.
DOI: 10.1007/978-3-319-44672-1_1

services extend reach of the health service provider, from traditional confines of care delivery to homes and hobbies, and from (doctor's) practice to everyday life.

This development trend, however, has a drawback which needs addressing before the services become the primary (if not even sole) way of conducting affairs in the health care domain, as has happened in the banking sector [3, 4]. In the core of health care has always been a patient whose preferences have formed a corner stone in health decision making. The advance of electronic health services is rapidly challenging this arrangement as 'traditional', face-to-face, health services are being replaced with their electronic counterparts, such as interactive virtual clinics [5] that employ web-based solutions in remote service delivery. In this kind of situation where electronic health services are becoming more and more commonplace, health care decision makers must ask themselves how the patients who prefer the traditional service models, or are simply unwilling to use technology, can be reached.

Unless this question is answered, we face a very real threat of placing a group of patients into the 'fringe' of health care as they are not able, or willing, to meet the demands of modern health service provisioning. In order to estimate how many people are in a risk of falling into the 'fringe', we need to look into the current statistics provided by the OECD[1]. First of all, in the OECD countries 80 % of individuals aged 16–74 were Internet users in 2013 compared to 60 % in 2005. In Luxembourg, Switzerland and Nordic countries, more than 90 % of the adult population were Internet users in 2013. As an opposite indicator, in Greece, Italy, Mexico and Turkey less than 60 % of adult population used the Internet. These differences are wider for older generations. For example, over 75 % of 55–74 year-olds in Denmark use the Internet while less than 10 % use Internet in Mexico. All in all, nearly half of the elderly people in the OECD countries used the Internet.

If we look deeper into consumer technology infrastructures, namely into broadband communication and wired band subscriptions, the penetration rate was 27 % in 2013. At the same time, wireless broadband subscriptions reached almost 75 %. As a related, more regional indicator, access to computers from home in the European Union (EU28) was 78.41 % in 2012 and the Internet access from home was at the same time 76.1 %. Interestingly, if we look into the discussed indicators, or to others related to the use or acquisition of ICT, two generalizations can be made. One, a penetration rate of 90 % or more is not common. For example, there are 34 Member Countries in OECD and only in one[2] country household broadband access was more than 90 % (Iceland, 92.4 %). Two, the development amongst 'high-tiered' countries, such as Iceland, tends to slow down with one exception; the amount of wireless broadband subscriptions. This will continue to rise with the expansion of Internet of Things (IoT) and Machine-to-Machine (M2M) communications, and wider availability of affordable mobile devices.

On the basis of this analysis a hypothesis can be made. In the near future, 10 % of the population in the OECD countries will be in a risk of falling into the 'fringe' as they are not able, or willing, to use the technology needed in the modern health service provisioning. If the risk is fulfilled, it will not only degrade the function of mature

---

[1] https://data.oecd.org/; http://www.oecd-ilibrary.org/ (Accessed: May 11, 2016).

[2] Data from South Korea was not comparable as it included mobile phone access.

health ecosystems that are turning digital; it will also prevent them from functioning in a sustainable fashion after the transformation. It follows from the nature of the problem, that the answer to the problem is not solely technological by nature as technology is a fundamental part of the problem. One possible answer is the use of a mediator ([6], p. 41), an intermediary in the crux of health, care and technology.

2 On Intermediaries and Mediators

There have been intermediaries in the field of health care for a long time. Patient advocates [7] and case managers [8] are practical examples of this role. Even the role of a practical nurse in home care is inherently intermediary, as it often involves interpreting health related information to the patient, and helping one in health related decision making. Most of these intermediaries are either formal positions or a 'job within a job'. As such, they are effectively subjected to the rules and regulations of the service providers. Individuals in these kinds of roles can be characterized as providerside intermediaries.

When it comes to more informal intermediaries, and ones working as patient-side intermediaries, the relatives often take up the task. In practical terms, this could mean that the spouse of an elderly person acts as a kind of case manager, ensuring that needs and demands of the elderly person are acknowledged. Even though these kind of informal intermediaries can become an expert in specifics of an ailment or condition (cf. [9], pp. 1809–1810), they are rarely competent in making formal and long-term health related decisions ([10], p. 180) such as, outlining a care plan. In practice, the lack of competence often follows from absence of formal training and education that gives perspective to health care professionals for handling a variety of conditions and ailments.

Mediators ([6], p. 41) are a specific kind of intermediaries. They operate on the patient-side, bridging the gap between the patient and health service providers. While mediator has similar characteristics to other intermediary roles in the field of health care, technology expands and differentiates the role from the others. More specifically, the role focuses on patient-side health technology and health care information systems, and acts as 'conduit' for technology and underlying services. As such, the role is a hybrid between those of a nurse and a technical support.

A real-world example on the need for a mediator comes from electronic health records (EHR), which are considered as essential tools for increasing collaboration between health care and social service providers, and patients ([11], p. 2). From the health service provider-side, these solutions contain functions relevant for daily clinical work, such as maintaining patient record and medication lists, tracking clinical tasks, and managing diagnostic tests. From patient-side, the functions are often limited to accessing patient record, and managing consent and authorizations. However, as the EU Digital Agenda states ([12], p. 29), the goal is set further than mere access to personal health information.

We are already seeing solutions which are more mature, not just in terms of technology, but also in terms of patient engagement. These solutions, such as the one used in Mayo Clinic, expand the reach of the traditional health care to everyday lives of the patients in the form of mobile health applications (cf. [13]). As these kinds of allegiances are becoming more and more common, diverse, and the traditional care becomes intertwined with aspects of well-being and fitness, a need for a mediator arises.

Today, not in some unforeseeable future, there are individuals who need someone to help them to bridge the gap between new and emerging technologies, and the new ways of providing care. In the basest form, this is realized as a need to fill in the forms that are online and not on paper anymore, or as a need to understand what the often jargon-filled health records actually stand for. Tomorrow, these same individuals may need help in conducting online medical examination in virtual clinics [5], or in uploading health-related information from their mobile health applications, which they are expected to use in managing a disease, such as diabetes.

3   Framework for Mediation

Basic division into provider-side and patient-side intermediaries is often enough as it conveys the essentials of the role and responsibilities. With this kind of dichotomy it is easy to understand (a) for whom the intermediary actually works for, and (b) what kind of legal and organizational constrains are in effect. However, in order to delve into intricacies of the actual position of an intermediary in relation to different stakeholders, a more specific framework is needed.

One way to categorize mediation is to look into activities of an intermediary; what is one's role in terms of mediated activities. In relation to the development and appropriation of new technologies, Stewart and Hyysalo [14] use a three-tiered approach that originates from Stewart's previous work [15] on the roles of cybercafés in the 1990s. In their work, they identify three distinct categories: (a) facilitating, (b) configuring, and (c) brokering.

Summarizing their work very briefly, facilitating can be described as providing opportunities to others. As such, it encompasses aspects of education, gathering and distributing resources, influencing regulations and setting local rules. Configuring, as the name suggests, refers to configuring technology but often only in a minor way as it also has a symbolic meaning. Configuring also refers to creation of space that facilitates appropriation (such as a cybercafé), influencing individual's goals and perceptions [14].

Lastly, brokering refers to a role intermediaries take when they set themselves up to support appropriation process. This includes negotiating on behalf of the represented individuals and institutions, for example when a new supplier is introduced to a project, or when requirements for a new product are discussed. Of the three categories, brokering can be seen as the most direct way of interaction between a user and a supplier [14].

The three-tiered approach devised by the authors [14] is befitting to the field of health care as it can be used for investigating activities of intermediaries already working in the field, and for analysing the missing ones. For example, work of a case manager in the field [8] is often related to brokering. This kind of an intermediary acts on behalf of a patient brokering services for the actual beneficiary. As material side of things is often present as well in the form of medical aids, medication or nursing supplies, facilitating is also an integral element in the work of an intermediary. Another aspect of facilitating that is commonly present in the role of an intermediary is education. For example, when a patient is discharged from a hospital to home care, patient education in the form of care instructions is often provided.

However, there is one particular area of facilitation that has not received sufficient attention; configuring. When it comes to electronic health care services, different health apps, or 'gadgets', users (patients) are often on their own. Or, if they need advice, users are expected to turn to online tutorials and support forums which themselves require a degree of technology literacy from the user. This kind of lack of configuring is not limited only to 'gadgetry' as it applies to a grander scheme of things.

As digitalization and concurrent service reform is changing the field of health care from the perspective of a user into a 'health space' [16] where information and services converge, a more profound need for configuring emerges. As the 'space' shifts and changes according the user's expectations and needs, configuring is needed to enable new uses (and users) of health-related information. As symbolic configuring touches upon issues such as rules and regulations on the use of health-related information, a specific area of conflict emerges.

It is commonly known that the health care domain is heavily regulated from the provider-side. However, on the user-side 'anarchy' reigns as the users tend to use technology according to their personal preferences, even moods. In this kind of contested landscape, born from personal control and self-service society, mediating a middle-ground between opposite views can be a challenge for any intermediary.

Another aspect related to intermediaries depicted by Stewart and Hyysalo [14] is related to the niche of a particular intermediary. While in the field of development and appropriation of new technologies, intermediaries may operate with a different focus. For example, if an intermediary operates solely on the design-side with a specific technology or product, the focus can be characterized as 'short and thing'. On the other hand, an intermediary who operates with a broad range of technologies or products with multiple actors (suppliers, end users, etc.) the focus can be characterized as 'long and fat'.

In the field of health care, formal intermediaries such as case managers [10] operate primarily on the provider-side, coordinating services on behalf of the actual beneficiary. What services are coordinated depends on the professional role of the intermediary, and on the prevailing health care system. For example, a case manager may operate solely on the field of mental health, and health service coverage varies from country to country [17]. This makes the reach of a formal intermediary a 'short' one, but the overall focus varies from 'thin' to 'fat' depending on the health care system.

Informal intermediaries, such as relatives who provide non-medical custodial care and assist the beneficiary in their everyday lives, often have a short reach as well. Reasons for this are often deeply embedded in the national legislation where managing affairs for someone else may be subject to legal or regulatory controls (cf. [18]). Another deeply embedded factor that effectively limits the reach of an intermediary is power imbalance present in the patient-physician relationship [19].

Even thought the relationship has changed (and it still is) from that of an age-old relationship between a priest and a supplicant, there are still barriers that prevent relationship from evolving into a more balanced one. Factors such as time, continuity of care, and even the facilities themselves still uphold the traditional (or paternalistic) patient-physician relationship [19]. In this kind of a setting, it is difficult to act as an intermediary with a 'long' reach. If the aim of the ongoing health care reform is to shift responsibilities related to care from physicians to patients, these barriers must be overcome; in mediation and otherwise.

4 Skills and Capabilities

It follows from the formulation of the role that in terms of relevant skills and formal training, there are two specific areas that need to be discussed when the focus is on patient-side intermediaries in the field of electronic health care services; health care and technology. In terms of health care, the mediators should possess a degree of skills relevant to the domain. These include skills in medicine and pharmaceuticals, and in particular, competence in terms of medical jargon. As the role is associated with health related decision making, a mediator should also have social acumen.

This implies that a mediator should be able give advice and answer questions related to the patient's health, and to be able to operate in emotionally charged situations with discretion. However, as the sphere of health care is expanding to adjacent fields, such as fitness, complementary therapies, and recreation, the most suitable skill set and the degree of skills, ultimately depends on the needs of the mediated person. However, regardless of these needs, the literacy skills are less subject for diversity. Of these, health literacy and eHealth literacy skills are a primary requisite for a mediator. In literature, these literacy skills have been defined in the following manner (example).

Health literacy
"The degree to which individuals have the capacity to obtain, process, and understand basic health information and services needed to make appropriate health decisions" [20].

eHealth Literacy
"The use of emerging information and communications technology to improve or enable health and health care" [21].
"The ability to seek, find, understand and appraise health information from electronic sources and apply knowledge gained to addressing or solving a health problem" [22].

Looking into these definitions provides a composite view to the problems associated with today's health information resources, and electronic health services. These two are of little use if the patients do not possess the sufficient skill to utilize them, and to analyse their relevance, applicability and degree of quality. As such, literacy skills are coupled with the notion of capability, and competence in them can be seen as a requisite for health care paradigm change commonly referred as empowerment.

Regardless of its vague nature (cf. [23]), the term empowerment captures well one of the current development trends in the field of health care. The term encompasses the change in the patients who are not willing to act as passive 'objects' and resign themselves to the hands of health service providers, but want to act as active 'subjects' and take matters of health and well-being into their own hands. In terms of this welcome change, mediators should be seen as enablers for the patients who consider technology as a barrier preventing them from achieving their aspirations.

5 Technology and Mediators

Technology has a dual role in relation to mediators; it is a part of the problem and solution as well. As a problem, technology creates a barrier between the patient, actual beneficiary, and modern health services, therefore creating a need for a mediator. As a solution, technology can be used by a mediator in fulfilling one's role. In the following, illustrative examples of this are provided.

Technology and Understanding. In terms of mediation understanding health-related information and its impact on the patient, are amongst the most important aspects of the role. This, however, does not come easily considering the state of today's health records. Even though parts of the records are structured and encoded according to the domain standards (CDA, LOINC, SNOMED, etc.), portions of the health records are in the form of free-form narrative. These portions often contain essential information in relation to patient's health and care, and as such their use in mediation is a necessity.

The portions, however, are often riddled with (a) domain-specific jargon, (b) abbreviations used in the service provider unit or field of specialty (for example, noradrenaline can be abbreviated as NA, NAd or norad), and (c) simple spelling mistakes (cf. [24], pp. 25–41). These factors alone have a negative impact on the quality of the free-form narratives. Another factor that needs attention in terms of language and understanding are the cross-border health care (cf. EU Directive 2011/24/EU) and health tourism, which are both increasing the number of potential health service providers.

In order to support care acquisition (especially from a foreign country), mediators should be able to understand health records regardless of (a) the language they are originally written, and (b) quality of the original health records. In this, natural language processing and information extraction technologies could provide a partial answer in the form of proofing and automated translation tools. These can be used in analysing the original documentation as is already done to a degree in the case of Bulgarian diabetic patients [25].

Another field of technology, which could be of use for mediator and for the mediated patient, is decision support aids. Especially when combined with health-related information from electronic health records, and consumer-side personal health records, technologies such as decision support scripts and expert systems can help a mediator in formulation of a care plan, in performing a virtual health check, or simply in forming an opinion to be presented to the patient. Especially now when numerous health information resources, and big data of the health care domain, are coming into a wider use, expert systems can very well be the key technology for aggregating and summarizing information from the often diverse and disparate sources.

Security and Privacy. In today's wired world, security and privacy are complicated issues. In order to use a specific application or service, it is possible that sensitive information must be accessed and distributed. When it comes to sensitivity, the most delicate issues are often associated with person's health, well-being or finance. In practice, this translates into person's medical record and payment history. Protecting this kind of information and enabling its safe use is not a simple matter. In addition to securing the information exchange and protecting health information sources there are other, primarily non-technical issues, such as consent and control that are of importance (cf. [26]).

Even though health and well-being are in the core of mediator's work, it does not automatically mean that a mediator should have open and all-inclusive access to patient's health, well-being or finance information. Some particulars of the information, such as current health status (such as, if the patient has diabetes), are often of the essence regardless of the mediator's sphere of operation. However, there are also certain particulars that are non-essential for mediation, or even harmful (for example, outdated medication lists or physiotherapy instructions). In order to ensure that mediators have access to the most relevant information, special attention should be put on information encapsulation (cf. [27], pp. 1329–1330; [6], p. 46).

In the context of this article, information encapsulation is an aggregate term which encompasses a subset of information management principles depicted by the Markle Foundation [17]. Of these principles (a) openness and transparency, (b) purpose specification, (c) collection limitation and data minimization, (d) use limitation, and (e) individual participation and control are incorporated into the term as they effectively depict the alignment of information between the patient and the mediators. The first principle of openness and transparency [26] is all about awareness; the patients should be able to know what information is collected about them, purpose of its use, who can access it, and where the information resides.

The second principle, purpose specification [26] is more instance-specific as it addresses the issue why the information is collected in the first place, or on each occasion of change of purpose. As an associated principle, collection limitation and data minimization [26], defines boundaries to the previous principle as it depicts the nature of information collection; information should be collected only for specific purposes and by lawful and fair means. Especially today, when unwarranted data collection by big technology companies is a common news topic, this principle has particular merit. In this, the principle of use limitation is in the same lines as it states that "personal information should not be disclosed, made available or otherwise used for purposes other than those specified" ([26], p. 4).

While the first principle of openness and transparency was about awareness, the last principle, individual participation and control, is about control. It defines [26] that the patients should be able to control access to their personal information, and review the used and stored information. Of the discussed principles, a subset of the ones presented by the Markle Foundation [26], the last principle outlines a very specific set of tools for mediation. With the control of access to the stored information, the patients are able to create specific aggregates, or subsets, of information for individual mediators to be used in their role.

It should be noticed that in particular the first and the last of the discussed principles contradict to a degree with the concept of mediation. As health related information is stored in multiform technological artefacts (such as EHRs), tools that are used for reviewing and limiting the use of information are often an integral part of the artefact. As these tools are technological by nature, they belong to the domain of a mediator. If a patient cannot operate these tools by oneself, without a mediator, how it is possible to maintain sensitivity in information encapsulation? In this, access definition and creation of information subsets on the level of existing aggregates (e.g. care record summary) or service providers (e.g. neurology polyclinic), could be of the essence.

6 How to Enable Mediation

Especially in the economic climate of today, enabling mediation is a challenge from the financial perspective. The public sector in most countries is struggling with the economic burden placed on them by increasing costs and decreasing income (e.g. tax revenue). It follows from this that it is difficult to justify costs of mediation, especially if they cumulate public health expenditure. Still, mediation as a function of home care is a viable option since some of the functions already overlap with mediation. For example, in the case of chronic heart failure, health promotion and teaching are often integral to the services provided at homes (cf., [28], pp. 1–2).

If the role of a mediator is seen as a 'job within a job' for a health care professional, such as a practical nurse, it creates a demand for advanced training in the field of health and technology literature in order to meet the demands of the role. As such, one way to understand the role and its alignment with the profession is to regard mediator as a specialist nurse in similar fashion to a critical care nurse or a school nurse. There are also specific fields of informatics in nursing which already incorporate elements of technology literacy, such as nursing informatics and telenursing. So basically, the role of a mediator has an established ground in the field of nursing.

However, with this approach the original characterization of a mediator as a patient-side intermediary becomes contested. If the role is enabled as a part of home care, funded by the state or the municipality, it essentially resides on a provider-side. One minor separating factor could emerge from the funding if mediation is enabled as a service provided by the private sector. As such a service, responsibilities of a mediator, and those of a professional working for a public health service provider, could remain separate (to a degree).

Another way to consider how mediation can be enabled is to regard mediator simply as a trusted person, a relative or acquaint. This is often the case with elderly persons who trust practicalities of their health-related endeavours into the hands of a spouse or a relative, and in the field of informal care where individuals providing assistance or care reside outside the framework of organized, paid and professional work. However, this interpretation is problematic as the informal mediator does not necessarily posses the required skills or capabilities. In this kind of a setting mediation does not necessarily base on domain expertise, but on anecdotal knowledge and second-hand experiences, as is often the case in layperson's health decision making (cf. [10]).

Possibility for a halfway solution emerges from multi-mediation; use of multiple mediators. Health-related endeavours that require a specific skill-set could be assigned to professional mediators, while endeavours which are less intensive in terms of skills and capabilities could be left to the relatives, spouses or other individuals working in the field of informal care. However, it should be stated that regardless of the composition of mediators, the self-determination of the patient should not be undermined.

When evaluating overall benefits, and even meaningfulness, of mediation, it should be understood that all benefits are not economic by nature. Instead of evaluating how much money has been saved in the form of reduced hospitalization, clinical outcomes and (other) quality indicators, such as quality of life, should be taken into account as well. Furthermore, indicators that are related to the efficiency and delegation of work are of interest in order to understand the effect of multi-mediation, which can be beneficial as well as disruptive. After all, there are no guarantees that the new ways automatically fare better than the traditional ones in terms of efficiency.

7   Conclusions

Health care, or more specifically health service provisioning, of tomorrow will differ fundamentally from that of today. Issues that have been part of the core values in the field, such as delegation of responsibilities between the doctor and the patient, are changing like never before. Electronic services are the harbinger of this change, which is not solely economic or societal by nature. As global economics are changing, so are the patients who, especially in developed countries, are 'wired from birth' as depicted by Brown ([29], p. 398).

Even though electronics services, and technology in general, can be regarded as belonging to the mainstream in most areas of business, there will always be individuals who are not able or willing to use them. Especially in the future, if electronic service gain similar foothold in health care, there is a risk that these kinds of individuals will fall into the 'fringe' of health services. In other words, they will not be in equal position when compared to other patients who possess a degree of technology (incl. health and eHealth) literacy skills.

In order to prevent this kind of development from occurring in the near-future, without negating benefits that can be gained from the use of technology in health service provisioning, we need intermediaries who act as representatives for those individuals who are at risk. These intermediaries, or domain-specifically mediators, are individuals who possess a specific set of skills that help them to operate in the crux of health, care, individual and technology.

Even today there are different intermediaries, such as patient advocates (cf. [7]), who operate in the field of health care on the service-provider side. The mediators, however, are individuals who operate on the patient-side, prioritizing the needs of the mediated patient before those of other actors (such as, health service providers or insurance companies) in relation to electronic health care services. It is the three factors of (a) patient (particularly interests of one), (b) health and care, and (c) technology that effectively set boundaries to mediation and to the role of a mediator.

Even though 'traditional', face-to-face services are still an option and in most cases the preferred way of providing care and associated consultation, there is a strong trend towards electronic services. In order to ensure that the health care of tomorrow is sustainable, and patients are treated equally in terms of service provisioning, we need to examine (a) what kind of intermediaries are needed, (b) what kind of mediation is really a viable option, and (c) how different actors are positioned in the electronic 'palette' that is the future of health care.

References

1. OECD: OECD Health Statistics, OECD Health Statistics 2014 (2014). http://www.oecd.org/els/health-systems/health-data.htm
2. Toiviainen, H.: Konsumerismi, potilaiden ja kuluttajien aktiivinen toiminta sekä: erityisesti lääkäreiden kokemukset ja näkemykset potilaista kuluttajina. Doctoral dissertation, University of Helsinki, Tutkimuksia 160, Stakes (2007)
3. Fox, S., Beier, J.: Online Banking 2006: Surfing to the Bank, Pew Internet & American Life Project (2006). http://www.pewinternet.org/files/old-media/Files/Reports/2006/PIP_Online_Banking_2006.pdf.pdf
4. Fox, S.: 51 % of U.S. Adults Bank Online, Pew Internet & American Life Project (2013). http://www.pewinternet.org/2013/08/07/51-of-u-s-adults-bank-online/
5. Krausz, M., Ward, J., Ramsey, D.: From telehealth to an interactive virtual clinic. In: Mucic, D., Hilty, D.M. (eds.) e-Mental Health, pp. 289–310. Springer International Publishing, Heidelberg (2016)
6. Lahtiranta, J.: New and emerging challenges of the ICT-mediated health and well-being services. TUCS dissertation, no. 176 (2014)
7. Mitchell, G.J., Bournes, D.A.: Nurse as patient advocate? In search of straight thinking. NSQ **13**(3), 204–209 (2000)
8. Mullahy, C.M.: Case management and managed care. In: Kongstvedt, P.R. (ed.) The Managed Care Handbook, 4th edn, pp. 371–401. Aspen Publishers, Maryland (2001)
9. Ballegaard, S.A., Hansen, T.R., Kyng, M.: Healthcare in everyday life - designing healthcare services for daily life. In: CHI 2008 Proceedings, 5–10 April 2008, Florence, Italy, pp. 1807–1816 (2008)
10. Ubel, P.A.: Patient decision making. In: Chang, A.E., Ganz, P.A., Hayes, D.F., Kinsella, T.J., Pass, H.I., Schiller, J.H., Stone, R.M., Strecher, V. (eds.) Oncology: An Evidence-Based Approach, pp. 177–183. Springer, New York (2006)
11. European Commission: Patient Access to Electronic Health Records, eHealth Stakeholder Group (eHSG) Report, June 2013 (2013)
12. European Commission: Communication from the Commission to the European Parliament, the Council, the European Economic and Social Committee and the Committee of the Regions - A Digital Agenda for Europe, Brussels, 19 May 2010 (2010)
13. Carr, D.F.: Apple Partners with Epic, Mayo Clinic for HealthKit, InformationWeek Health Care, 3 June 2014 (2014)
14. Stewart, J., Hyysalo, S.: Intermediaries users and social learning in technological innovation. Int. J. Innov. Manag. **12**(3), 295–325 (2008)
15. Stewart, J.: Cafematics: the cybercafe and the community. In: Gurstein, M. (ed.) Community Informatics, pp. 320–338. Idea Group Publishing, Hershey (2000)
16. Lahtiranta, J., Koskinen, J.S.S., Knaapi-Junnila, S., Nurminen, M.: Sensemaking in the personal health space. Inf. Technol. People **28**(4), 790–805 (2015)
17. World Health Organization: World Health Statistics (2015)
18. Eika, K.H., Kjølsrød, L.: The difference in principle between the poorly informed and the powerless: a call for contestable authority. Nord. Soc. Work Res. **3**(1), 78–93 (2013)
19. Joseph-Williams, N., Elwyn, G., Edwards, A.: Knowledge is not power for patients: a systematic review and thematic synthesis of patient-reported barriers and facilitators to shared decision making. Patient Educ. Couns. **94**, 291–309 (2014)
20. Ratzan, S.C., Parker, R.M.: Introduction. In: Seldon, C.R., Zorn, M., Ratzan, S.C., Parker, R.M. (eds.) Current Bibliographies in Medicine 2000-1: Health Literacy. National Institutes of Health, National Library of Medicine, US Department of Health and Human Services, Washington (2000)

21. Eng, T.R.: The e-Health Landscape: A Terrain Map of Emerging Information and Communication Technologies in Health and Health Care. The Robert Wood Johnson Foundation, Princeton (2001)
22. Norman, C.D., Skinner, H.A.: eHealth literacy: essential skills for consumer health in a networked world. JMIR **8**(2) (2006). http://www.jmir.org/2006/2/e9/v8i2e9
23. Aujoulat, A., d'Hoore, W., Deccache, A.: Patient empowerment in theory and practice: polysemy or cacophony? PEC **66**, 13–20 (2007)
24. Suominen, H.: Machine learning and clinical text: supporting health information flow. TUCS dissertation, no. 125 (2009)
25. Nikolova, I., Tcharaktchiev, D., Boytcheva, S., Angelov, Z., Angelova, G.: Applying language technologies on healthcare patient records for better treatment of Bulgarian diabetic patients. In: Agre, G., Hitzler, P., Krisnadhi, A.A., Kuznetsov, S.O. (eds.) AIMSA 2014. LNCS, vol. 8722, pp. 92–103. Springer, Heidelberg (2014)
26. Markle Foundation: Common Framework for Networked Personal Health Information: Overview and Principles, Markle Foundation (2013)
27. Denley, I., Smith, S.W.: Privacy in clinical information systems in secondary care. BMJ **318**, 1328–1331 (1999)
28. Fergenbaum, J., Bermingham, S., Krahn, M., Alter, D., Demers, C.: Care in the home for the management of chronic heart failure, systematic review and cost-effectiveness analysis. J. Cardiovasc. Nurs. **30**, S44–51 (2015)
29. Brown, S.A.: Household technology adoption, use, and impacts: past, present, and future. Inf. Syst. Front. **10**(4), 397–402 (2008)

Tactile Maps - Safety and Usability

Stina Ojala[1($\boxtimes$)], Riitta Lahtinen[2,3], and Helinä Hirn[4]

[1] Department of Information Technology, University of Turku, Turku, Finland
stina.ojala@utu.fi
[2] Finnish Deafblind Association, Helsinki, Finland
riitta.lahtinen@kuurosokeat.fi
[3] ISE Research Group, University of Helsinki, Helsinki, Finland
[4] O&M Consultant, Private Entrepreneur, Helsinki, Finland
helinahirn@gmail.com

Abstract. Maps are regularly used in planning a trip or a route, e.g. when going on a holiday abroad or driving to a meeting in a place one has not visited before. Sighted people can get hold to a map quite easily as they are available both in paper copies and furthermore virtual maps are more and more widely available too.

When one is visiting a city abroad one uses maybe more than one map at a time - city map as well as a route map of the public transport. These can be carried quite easily, as well as referred to on a regular basis. There are also maps of bigger venues like seating plans of auditoriums or concert venues. For the visually impaired tactile maps are also a safety issue [1].

Keywords: Tactile maps · Visual impairment · Mobility · Safety

1 Introduction

Maps are used to learn about an area and planning routes within it [2]. Maps are also used in education to teach about the awareness of the world in general as well as in more detail. The visual maps are easily reproduced and hand-held while tactile maps are more expensive and not as easily referred to. A map is a physical array of symbols to represent and depict the spatial relations of objects in a physical environment [3]. It is a two-dimensional representation of the environment showing specific features of the terrain in terms of their relative size and position [4]. The design of tactile maps generally follows the same guidelines as those used in visual maps, but the basic image must be edited for tactile exploring [5–7]. A tactile map cannot be a mere translation of visual information into tactual form. A tactile map, in other words, is a raised, specially-adapted tactile representation of the specific cartographical location in question. Both visual and tactile maps are concrete thus they are different from verbal maps and description.

How the map feels is more important in a tactile map than how it looks [5, 8, 9]. The important aspects to be considered when preparing tactile maps are according to [8]: 'the ability to discriminate lines, textures, size, labelling, and use of colour'. Tactile maps can give a greater spatial understanding than either a direct experience of moving

© Springer International Publishing Switzerland 2016
H. Li et al. (Eds.): WIS 2016, CCIS 636, pp. 15–22, 2016.
DOI: 10.1007/978-3-319-44672-1_2

through the environment [10] or a direct experience supplemented with verbal explanation [11]. The relevant information is presented clearly in a map with relative
simultaneity, and without other difficulties associated with travel in the real
environment.

It would be essential to create coherent guidelines for tactile map design. The
guidelines should take into account the differences in age, vision and other abilities of
the potential users. The design should include the map size and format, the choice of
symbols and the scale [12]. If this were achieved, it would be possible to have globally
standardised maps.

Haptic exploration in reading tactile maps uses the combination of information
from hands, movement and sense of touch. When a person touches an object it is at a
certain orientation to him/her. It is a relation in space between a person and the object
being touched. Haptic exploration includes information about the orientation, tactile,
movement though tendons and muscles in the hand. It is a combination of these bits of
information [13] among others. You have to have a physical contact to the object in
order to gather information by haptic exploration. On a map hands recognise the
temperature of it on first contact, but on exploring by movement one gathers information of the shape, size, height and material as well as the alterations on the map [14].
With haptic exploration one processes the details, differences and relations in the
map. The process involves also the combination of these information patterns into a
mental map, the connections between map and real world environment and decipher
the details for safe movement within a surrounding.

According to [15] the muscle and proprioception senses are more important in
movement and receive the environment as visual pathways are not available. The
differences in materials and surfaces are felt through the muscles and proprioception in
fine detail. The basic orientation system helps with the mental maps too.

In this paper we present examples of tactile maps manufactured using different
types of materials and on some of them we have user perspectives on their usability.
The maps in the article are currently used in O&M consulting with both blind and low
vision people. Gathering end user comments on usability and importance of maps is,
we think, the preferable way in trying to develop international standards for tactile map
manufacturers. There are differences between the opinions of blind and low vision end
users of tactile maps and we present both views when obtainable.

2 Reading Maps

The conventional method of direct familiarisation with an area by travelling through it
can be time-consuming. When a child who is blind explores an unfamiliar environment
he or she can study a tactile map before entering the area. It allows the child to explore
the area independently, thus avoiding the situation of having all the information coming
through another person. Even a brief exposure to the map is an effective and adaptable
means for introducing the spatial structure of a novel area [16].

In 1970s [17] compared the performances of good and poor map readers, who were
school-age Braille readers, in order to distinguish the necessary elements for effective
map exploring procedures. Continuous movement in line tracing, the ability to search

for shapes, recognition of shapes, comparison and differentiation of shapes, and locating distinctive features on the maps were the important factors influencing good performance, as well as the recognition and tracing of shapes which were juxtaposed to other shapes.

Good map readers searched the map completely by using one finger instead of the flat of the hand or several fingers. They also picked out a point of origin and traced around the shape in a continuous motion and returned to the point of origin; they did not search the area between the lines or contours. [18] found that children who were good tactile map readers had strategies of the same kind as sighted children.

One research group [19] videotaped the performance of 16 adults with visual impairment as they explored tactile maps. The subjects used their hands singly or together, and they could be fixed or move across the map. Sometimes, the hands were held flat, and the palm was used to gain an overview. The fingertips were clearly the most important in tactile map reading. Often one finger was used for fine discrimination of raised details and four fingers for general exploration. Subjects used their fingers constantly to move over the surface of the map; at other times the fingers repetitively traced back and forth along a symbol, or they traced a line. Good readers employed a mix of single finger, multiple fingers and whole hand-based techniques. Usually the map is first scanned with both hands. It is done from top to bottom to get the size of the map. The studying of the map legend symbols and Braille on the map is done with fingertips as the fingertips are most sensitive in details [20].

3 Factors Affecting Readability of Maps

The factors affecting readability of maps differ in visual and tactile maps. Visual map readability is affected by choice of colours, scale and abundance of information in the map. Also overall size matters—whether the map is easily folded and easy to track a certain path in a map that stretches several pages. The colours can be given a specific meaning that needs to be explained in a map legend.

In tactile maps readability is affected in two different scopes: tactile and low-vision readability. Often tactile maps are designed for low vision users as well. Here strong visual contrasts are important as well as the orientation of the map concerning lighting conditions. Tactile readability is affected by abundance of information—too much information results in low readability. Orientation of the map affects readability as well. While visual maps can be easily used outside as well as inside in tactile maps weather conditions affect readability of them. This is more poignant in the north where snow and cold weather can hinder readability of the map drastically.

In tactile maps the orientation of the map in the surrounding can greatly affect its readability in terms of physiological restrictions in using one's hands. That is a map that is upright on the wall can be less readable because of the restrictions in flexing muscles of the wrists. In maps on the wall one's height affects the maps readability as one cannot reach but to a certain height on the map. This is more important and has to be more widely studied and noted when erecting tactile maps. Reading a map on the wall is not the best for ergonomy. On the other hand, if a map is erected horizontally there is a possibility people start using it as a table surface thus using it for coffee cups, etc. That means maps should be erected inclined to prevent its use as a table surface.

Desirable characteristics of tactile maps are durability, sharpness of border lines, surface texture, recognisable symbols, and availability. Maps can be located and available in various places, and they should withstand abrasive use, chemical exposure, and adverse weather conditions. At the same time they should be pleasant to touch, consistent in symbol representation and they should have distinguishable lines to trace.

4 Examples of Tactile Maps

In this section we present examples of tactile maps with discussion or end-user opinions where available. There are four different solutions of tactile maps, both inside and outside. These maps in question are unique and unusual as the materials in them are not traditional but the choice of materials has been influenced by their location and the vision of the architect. That is to say these maps are not mass produced but rather individually designed into their place.

Wherever in the world there is a tactile map, the processing and reading of the map serves exactly the same purpose: to orientate oneself in the surroundings and to find out the routes within, to familiarise oneself with the environment. The context is always the same, to process the information from the map and map it into the environment. When a person with a visual impairment reads a map, the reading process starts with the map legend, the compass rose if available, then the general framework before starting to explore the details in the map. Studying the map is quicker when studying it with an O&M instructor or a sighted guide. Braille is used in tactile maps all over the world.

Paris metro tactile map is placed near a wall on waist height. It is erected inclined. There is no guiding to the map with guiding tactile strips, but the route needs to be memorised. The material is metal, it is dark and with no colour alterations. It is difficult to use with low vision because of lack of colour contrasts. It has a good tactile feel, but the metal is slightly cold to touch. The map illustrates different platform areas and exits (Fig. 1).

The tactile map of an institute for the visually impaired in Finland (Keskuspuisto Vocational College) is erected outside of the building, in a taxi waiting area (Fig. 2). The map includes the premises of the institute as well as routes in the immediate vicinity of the institute. There are the connections to public transport nearby (buses and train stations). The map has colours but they are low in contrast. It might be partly due to wear and tear as the map is outside and the fact that it is subjected to adverse weather too, especially in the wintertime.

Kamppi bus terminal map (Fig. 3) is ceramic so has good tactile contrasts as well as a good contrast in colours, which increases low vision readability. Each of the floors in the centre is presented in a separate map, which also adds both low vision and tactile readability. It displays bus terminal functions and exits, so it promotes route planning and presents a general overview of the layout of the centre.

The tactile map of Iiris, the service and activity centre for the visually impaired in Finland (Fig. 4), has original choice of materials including wood, which makes it beautiful both visual and tactile sense, it is smooth to touch. It however makes the

Fig. 1. Paris metro tactile map

Fig. 2. Keskuspuisto Vocational College tactile map

Fig. 3. Kamppi bus terminal tactile map

Fig. 4. Iiris centre tactile map about here

updating a challenge. The map has general overview of all the floors of the centre with the functions as well as the guiding tactile strips depicted. However, due to challenging updating procedure the map is not in active use as it refers to functions that are no longer present.

5 Conclusions and Future Vistas

There is no systematic convention about the place of the map legend on a tactile map but it depends on the manufacturer of the map. For example in Finland there are maps which have map legend information scattered to different places on the map. This decreases map's readability. Another issue is that there are no conventions on the symbols on the map, in other words the symbols differ from one map to another. There is a distinct need for a standardised, international set of symbols to be used in tactile maps world wide. It is also utterly important that the map is placed in such a way that it can be found easily and the place is calm enough for a person to stay there to study it.

Often in construction phase the architect wants a certain scheme to the building and that also includes the visual design of the tactile map. However, this is not the optimal solution for a tactile map as there are certain features that differ from visual to the tactile perception. Tactile maps should not be judged by visual standards but according to the characteristics of the sense of touch and how does the map feel instead of how it looks like. For example the map in Iiris (Fig. 4) was designed as a work of art but it resulted in updating being a challenge. To have a map correctly updated is also a safety feature [1].

References

1. Lahtinen, R., Palmer, R., Ojala, S.: Practice-oriented safety procedures in work environment with visually and hearing impaired colleagues. In: Saranto, K., Castrén, M., Kuusela, T., Hyrynsalmi, S., Ojala, S. (eds.) WIS 2014. CCIS, vol. 450, pp. 109–119. Springer, Heidelberg (2014)
2. Parente, P., Bishop, G.: BATS: the blind audio tactile mapping system. In: Proceedings of the ACM Southeast Regional Conference, pp. 132–137 (2003)
3. Landau, B.: Early maps as an unlearned ability. Cognition **22**, 201–223 (1986)
4. Thrower, N.: Maps and Civilization: Cartography in Society (245). University of Chicago Press, Chicago (1996)
5. Eriksson, Y.: Att känna bilder [To feel pictures], pp. 84–87. SIH Läromedel, Solna (1997)
6. Eriksson, Y., Wollter, S.: Från föremål til taktil bild [Converting the object into a tactile picture], pp. 31–41. Tryck Nya Tryckeriet, Lyckelse (1997)
7. Gardiner, A., Perkins, P.: 'It's a sort of echo…': sensory perception of the environment as an aid to tactile map design. Br. J. Vis. Impairment **23**(2), 84–91 (2005)
8. Berlá, E.P., Butterfield, L.H., Murr, M.J.: Tactual reading of political maps by blind students: a videomatic behavioral analysis. J. Spec. Educ. **10**(3), 263–276 (1976)
9. Andrews, S.K.: Applications of a cartographic communication model to tactual map design. Am. Cartographer **15**(2), 183–195 (1988)
10. Ochaíta, E., Huertas, J.A.: Spatial representation by persons who are blind: a study of the effects of learning and development. J. Vis. Impairment Blindness **87**(2), 37–41 (1993)

11. Ungar, S., Blades, M., Spencer, C.: The use of tactile maps to aid navigation by blind and visually impaired people in unfamiliar urban environments. In: Proceedings of the Royal Institute of Navigation, Orientation and Navigation Conference 1997. Royal Institute of Navigation, Oxford (1997)

12. Rowell, J., Ungar, S.: The world of touch: an international survey of tactile maps. Part 2: design. Br. J. Vis. Impairment 21, 105–110 (2003)

13. Gibson, J.J.: The Senses Considered as Perceptual Systems. Waveland Press/Houghton Mifflin Co., Prospect Heights/Boston (1966)

14. Klatzky, R.L., Lederman, S.J.: Toward a computational model of constraint-driven exploration and haptic object identification. Perception 22, 597–621 (1993)

15. Jokiniemi, J.: Kaupunki kaikille aisteille. Moniaistisuus ja saavutettavuus rakennetussa ympäristössä. Teknillinen korkeakoulu, Arkkitehtiosasto, Kaupunkisuunnittelu, Espoo (2007)

16. Hirn, H.: Pre-maps: an educational programme for reading tactile maps. Academic dissertation, University of Helsinki (2009)

17. Berlá, E.P., Butterfield, L.H., Murr, M.J.: Tactual reading of political maps by blind students a videomatic behavioral analysis. J. Spec. Educ. 10(3), 265–276 (1976)

18. Ungar, S., Blades, M., Spencer, C.: Visually impaired children's strategies for memorizing a map. Br. J. Vis. Impairment 3(1), 27–32 (1995)

19. Perkins, C., Gardiner, A.: Real world map reading strategies. Cartographic J. 40(3), 265–268 (2003)

20. Gardner, E.P., Kandel, E.R.: Touch. In: Kandel, E.R., Schwartz, J.J., Jessell, T.M. (eds.) Principles of Neural Sciences. McGraw-Hill, New York (2000)

The Impact of Digitalization
on the Medical Value Network

Teijo Peltoniemi[⊠]

University of Turku, Turku, Finland
teijo.peltoniemi@utu.fi

Abstract. Digitalization equalizes information asymmetries which increases economic efficiency and transforms many lines of business. It can be argued that digitalization can do the same in the medical and health care market. This article conceptualizes the medical and health care value network and its digitalization utilizing concepts of information asymmetry and value networks. Inefficiencies within the network relating to incomplete information and information asymmetry are identified. Some digital solutions to these issues are suggested including ePrescription and the automatic medicine dispenser.

Keywords: Digitalization · Medicine · ePrescription · Information asymmetry · Value network

1 Introduction

Digitalization has been a buzzword in the 2010s. It has been defined in many ways. In its simplest form it has referred solely to replacing a manual process with an electronic process. An example of digitization is when an organization replaces a paper document with an electronic one. Digitalization on the other hand has been defined as a more far-reaching phenomenon where an organization changes its structures and ways of working [1]. Terms such as re-engineering have been associated with digitalization as well as with an ever growing amount of data and information.

The key components of digitalization are the global digital network and cheap, commoditized computing resources [2]. Information is created at an increasing pace and wide information resources have become available to the masses. This has changed many lines of business. In this article we will examine the potential of digitalization in health care and the medication market. The main research questions we intend to answer are specifically:

- How incomplete information and information asymmetries in the medical value network can be managed?
- How digital tools can support this?

© Springer International Publishing Switzerland 2016
H. Li et al. (Eds.): WIS 2016, CCIS 636, pp. 23–36, 2016.
DOI: 10.1007/978-3-319-44672-1_3

2 Theoretical Background

2.1 Value Networks

Digitalization can be viewed through the concept of the value network. The value network refers to the network in which an organization operates with other organizations with the aim of producing value for the end-customer in the form of products and services. The value network concept focuses on the information flows between the actors in the network and regards information as a key factor for gaining competitive advantage [3].

The nodes, actors, in the network include customers, outsourcing partners, sales partners and all other relevant parties who collaborate in one way or another with the organization. The key actor in the network is the customer for whom the network creates value in the form of products and services.

The value network is a broader concept than that of the supply chain. Where the supply chain relates mainly to the operative level of an organization and its linear physical processes the value network is a strategic concept. The value network is an emerging structure changing its shape frequently [4].

Each of the actors in the value network possess different assets. These can be tangible or intagible [5]. One intangible asset is information which is utilized within the network as it operates. For example there is information about customers, products, manufactured quantities, prices and so forth.

Information can be scarce and hence highly priced. Actors possessing valuable information have constituted powerful forces in their value networks. Digitalization is changing this however. Information has increasingly become freely accessible and obtaining information has become cheap [6]. This equalizes the power balance within the value network and changes the dynamics in many lines of business, arguably also in those relating to health care and medication [7].

2.2 Asymmetric Information

Akerlof [8] famously discusses the concept of information asymmetry in his seminal work. This refers to a transaction event in which one party has more information than the other. The exploitation of this unequal situation for the benefit of the one party leads to a market failure. Akerlof studied the market of used cars and showed that fraudulent car sellers selling bad cars, called lemons, caused good used cars to disappear from the market. This is called adverse selection.

Information asymmetry does not only occur when the seller has more information. Related concepts include moral hazard which refers to a situation where party A has to bear the risk for party B's actions but does not have adequate information on party B's intentions nor ways to ensure that party B will perform in an efficient manner. Examples of this include when an insured person takes unnecessary risks or when a taxpayer hides information on his or her income from the tax authorities and neglects to pay taxes.

The health care market is subject to information asymmetry on a wide scale as Arrow [9] discusses in his well known article. This is manifested in many ways: for instance, the patient has significantly less information on medical treatments than the physician. The physician on the other hand cannot be certain that patients will follow the prescriptions suggested to them. The information on the past performance of a physician is also scarce which complicates the purchase of medical treatment. Furthermore, the efficacy of a treatment is uncertain at most times.

It can be argued that digitalization can help overcome issues with information asymmetry [6, 7]. As discussed above digitalization adds information which was previously scarce or non-accessible and makes this available to new actors. This can be seen as reducing information asymmetry. The amount of available information is increasing at a fast pace in many fields. What is the situation in the healthcare and medical market? It can be argued that information asymmetry is still a significant problem which could be alleviated by digitalization.

3 Methodology

This is an exploratory study based on existing literature. This is supported by one interview that was used to set the scene in the first place. The focus is on forming an overview of the context and propose future research to gain more profound insight into the area. The context is further explored through illustrative examples.

Topics that we are discussing in this article, such as ePrescription and other ICT-based innovations in the area of medicine, are forming the research area for the writer, which inevitably affected their inclusion into the analysis. The perspective how they are analyzed however is new and we explore the literature in order to find answers for the research questions set in the end of Sect. 1.

As this is an exploratory study we do not intend to answer conclusively to a clearly bounded research questions. Instead, we intend to outline a new way to conceptualize a phenomenon, namely digitalization within the medical area. We acknowledge that further study is certainly needed to inspect and fully understand the phenomenon.

The research followed an iterative process, which is typical with qualitative studies [10]. At the beginning of the study a representative of the Finnish pharmaceutical industry was interviewed [11] in order to identify and outline the problem area. An open interview technique was utilized which was necessary at that stage as the exact research questions were yet to be defined.

The obvious shortcoming of this approach is that it presents only a single perspective to a complex environment. This however should be acceptable as the purpose of the interview was merely to explore the problem area, and not to draw any conclusions to the research questions. Based on the interview, writer's previously gained knowledge in the field, and literature, a model for the health care and medical market was conceptualized utilizing the value network concept.

In the next phase of the study targeted queries were ran to find relevant and recent articles on the management of incomplete information and information asymmetry in identified parts of the medical supply chain. We did not follow a full systematic literature research method, as it was not practically feasible. The reason for this is that the puporse was to explore the problem area through multiple illustrative examples (ePrescription and other ICT-based innovations). Databases such as PubMed, ScienceDirect, ACM and ProQuest and public search services such as scholar.google.com were utilized.

It also became clear that articles combining the discussed technologies and information asymmetry are scarce. We therefore had to study materials in iterative manner and refine the queries in the course of the study after identifying new potential key words. We however acknowledge that the different topics deserve own dedicated studies and these should be supported by systematic literature research.

4   The Medical Value Network and Information Asymmetries

One of the key problems with health care systems is fragmentation [12]. Dispersed organizational structures with poor information flows can lead to higher costs and poorer quality of care [13]. This makes a network and customer centric approach such as the value network concept particularly attractive as an analysis tool. In the case of the medical market it leads to the creation of understanding on how organizations such as service providers, physicians and pharmacies are linked together and how they co-operate in order to produce value for the patient. Furthermore, it enables the identification of bottlenecks and information gaps which impact a network's performance.

The key segments and actors in the health care value chain can be specified as follows [14]:

- Payers: e.g. government, employers and individuals (patients).
- Fiscal intermediaries: e.g. insurers.
- Providers: e.g. hospitals (service providers), physicians and pharmacies.
- Purchasers: e.g. wholesalers.
- Producers: e.g. medicine and medical device manufacturers (pharmaceutical companies).

Figure 1 illustrates the medical value network populated with these actors. The diagram is intended merely for illustrative purposes and it is not an exhaustive model of the market. Associated regulatory and professional institutions have been omitted for the sake of simplicity for instance. It is a top-level view and does not drill into the details of the diverse ecosystems surrounding the pharmaceutical industry or service production. The purpose of the diagram is to bring forward the conceptual framework for digitalization within the medical care environment.

The dots in the diagram denote actors. The arches illustrate relationships between actors. A physician is employed by a service provider for instance and prescribes treatment for a patient who complies or does not comply with the prescription.

If we consider the information assets which different actors possess we can intuitively draw some findings:

- A physician has information about treatments.
- A patient has information concerning his or her personal health and medication history.

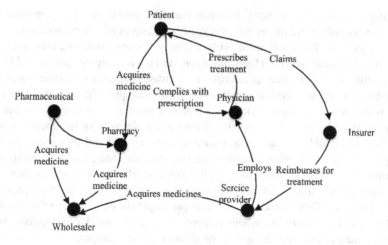

Fig. 1. Medical value network

- A patient has information on his or her compliance as regards a prescribed treatment.
- A pharmacy has information on medicines, alternatives and their market prices.

Considering these information assets and examining the network we can identify a number of information asymmetries between actors. For example:

- Physician – patient: a patient does not have information on the past performance of a physician nor can the patient be certain the care prescribed is effective.
- Patient – physician: a physician cannot be certain the patient complies with the prescription suggested for them.
- Patient – insurance firm: an insurer does not have full information on the actions a patient takes nor the means to monitor them.
- Patient – pharmacy: a pharmacy does not have full information on all past prescriptions ordered for a patient.
- Pharmacy – patient: a patient rarely has extensive information on medicines or on their prices in order to be able to accurately assess or compare the pricing of particular courses of treatment.
- Service provider – medicine wholesalers: medicine spoilage is typically a significant source of costs [15]. It can be caused by the lack of demand management related information.

5 Decreasing Information Asymmetry Through Digital Solutions

5.1 Electronic Prescription

Electronic prescription, or ePrescription, can be defined as "the use of computing devices to enter, modify, review, and output or communicate drug prescriptions" [16].

Regarding information assymetry between patients, physicians and pharmacies the power of electronic prescription lies specifically in improved communication.

The impact of ePrescription on prescription errors and increased communication within the healthcare service chain has been studied in different accounts [17–20]. Various studies suggest that ePrescription decreases medication errors caused by unclear handwriting or otherwise unclear or incomplete prescriptions. Another major benefit is that a more complete view of a patient's medication history is generated.

This increased information can be utilized in the delivery of healthcare services. The prescriber should have adequate information on current and past prescriptions to avoid prescribing overlapping or otherwise non-suitable medication for instance and ePrescription shows the potential to make this possible. ePrescription also undermines a patient's potential to manipulate the information asymmetry associated with prescribed drug misuse cases. This is since all previous prescriptions are visible to the physician. The physician can determine whether there are existing valid prescriptions before prescribing medication such as opioids or stimulants for example.

Lizano-Díez et al. suggests that polymedicated patients specifically have benefited in terms of decreased prescriptions and hence medication costs after the launch of the ePrescription system [21]. The key factor is the increased communication between pharmacies and prescribers which prevents medication errors as well as overlapping prescriptions for medication.

The Finnish Patient Data Repository (KanTa) is a database containing personal healthcare records, including prescriptions, collected from healthcare service providers' patient information systems [22]. It allows a physician to examine thoroughly the healthcare history of a visiting patient when specifying a course of treatment. This was not always possible previously since information was not always shared between different service providers and hence the physician might have lacked information on past prescriptions ordered by another service provider.

The wider debate on the ownership of patient information is relevant with reference to ePrescription. Broadly speaking the question is whether patients should own their own data records. According to the current practice and legislation service providers own healthcare records which they maintain [23]. This field however is emerging and will have many applications; a patient can allow access to his or her personal data in exchange for economic benefits for example [24].

5.2   The Generic Substitution System

The generic substitution system of medicine refers to a system which enforces the use of generic medication over brand name medication. The implementation of this system is possible at many stages of medicine delivery. Physicians can look for the most inexpensive medication when issuing prescriptions for example. Pharmacies can also suggest a generic medicine if one is available. In Finland the latter has been enforced by law; pharmacies are obliged, with some exceptions, to suggest a low cost alternative if one is available [25].

Generic substitution can be regulated by legislation, as is the case in Finland, where the generic substitution system is accompanied by the medicine reference price system [26]. Pharmacies have to follow reference pricing and it is easy to conclude that the market is highly regulated.

There is evidence that both the generic substitution and reference price systems have had an impact on medicine prices in Finland [27]. The growth of gross pharmaceutical sales has slowed in the 2000s suggesting that these have had an effect [28]. In the US it is estimated that wider adoption of generic substitution would save insurers and patients $9 billion annually in outpatient care [29]. Generic products cost $45 less on average so the impact would be significant for patients and other payers [29].

Regulations and related institutions are typically seen as a way to decrease information asymmetry in a market. It is for instance unlikely that a consumer can assess medicine prices as well as a professional nor do they have the capability to assess a particular substitute for a prescription so the high level of regulation can be considered justified in the medicine delivery market.

There are also disadvantages to regulation. Regulation is expensive to implement and maintain and the institutions involved generate costs. Regulation typically does not keep pace with technological innovations [23]. Often regulations can form barriers to the adoption of new innovations and remain as a subject of continuing lobby by parties who have an interest in keeping the market in its current state.

Digitalization has introduced new trust mechanisms which has undermined the need for regulations [6]. These are typically implemented in platforms which host equal information about suppliers, buyers and services for all platform users. Digitalization has also improved the management of information asymmetry relating to generic substitution: various digital services have emerged which support the search for generic substitute medicines (e.g. [30]). They can integrate open data sources such as pharmaceutical databases and can be used by both physicians and patients.

A key barrier to generic substitution which cannot be overcome by public policies other than increased education is the lack of knowledge amongst both physicians and patients [29]. Physicians often remain cautious in their approach to generic medicines as do patients who often associate higher price with higher value.

Stenner et al. [29] suggest that ePrescribing decision support tools can provide the means to close this knowledge gap. According to their study enhancing the ePrescription system with these mechanisms increased the uptake of generic prescriptions.

5.3   The Automatic Medicine Dispenser

Patient adherence is a key factor in the successful delivery of medication [31]. This refers to a patient's compliance with a prescribed treatment such as medication. Non-compliance with prescriptions costs the US healthcare system $300 billion annually [32]. Non-adherence can be intentional or unintentional [33] and it is problematic from the physician's point of view: the physician can only trust the patient complies with the prescription and follows the treatment.

When assessing a returning patient or monitoring a patient's recovery information on adherence has traditionally been incomplete. Information asymmetry follows when a patient hides information from a physician, intentionally or unintentionally. This can be managed with automatic medicine dispensers.

Modern automatic medicine dispensers are connected online and raise alerts when a patient discards a prescribed course of medication [34]. This information can be provided to the physician in real time and consequently the patient has less opportunity to disguise or deny non-compliance. Although the benefits have been accepted automatic medicine dispensers have yet to be widely adopted, particularly in home use settings.

The reasons for non-adoption have been studied in Finland and they include costs and regulations [35]. Dispensers are typically expensive and they are not fully covered as part of standard health plans. They also need to incorporate failsafes and assurance mechanisms which makes them expensive to build. Regulations also hinder adoption since the fulfilment and validation process requires pharmacy visits.

5.4   Outcomes Based Medicine and Continuous Health Monitoring

The power relationship between the healthcare expert and the patient has traditionally been highly unequal. This is because the expert holds significantly more information on treatments. This information asymmetry can be manipulated for the furtherance of the expert's own interest [36]. For instance if a service provider's revenues are based solely on patient volumes and the efficacy of prescribed treatments is not monitored the quality of care can be compromised to accommodate increased patient flow and the resulting revenues.

We have discussed means for managing asymmetric information in previous sections, namely ePrescription, generic substitution and the automatic medicine dispenser. However these all have shortcomings. Intentional non-adherence can be difficult to control should a patient deliberately wish to disregard a particular medication. On the other hand there is no guarantee that a prescribed medication specified by a physician is the best available treatment for a particular condition.

Outcomes and quality based approaches have emerged to tackle this. They are based on incentivising healthcare providers on the outcomes of treatment. There is evidence that this approach has positive effects on the quality of care [37] and it is actually not an unusual goal for healthcare systems worldwide [12]. These approaches are based on a long-term relationship between physician and patient and on continuous monitoring of medical treatment.

Value-based methods can be utilized to manage information asymmetries in care and medication delivery. Firstly these methods produce information about the efficacy of a prescribed treatment. When recovery is monitored on continuous basis it is possible to assess how well a treatment works for a given patient. This can be linked to incentives and reimbursements associated with the treatment. As a result it equalizes the power relationship between the physician and patient.

Secondly this approach can also help with another instance of imbalance of information between these actors. If the incentive the physician receives depends on the efficacy of the treatment they prescribe then compliance with particular prescriptions should be assured. As recovery is monitored continuously the physician obtains more information regarding the patient's health. This information should allow the physician to determine the extent to which the patient is complying. This can be used to manage the associated moral hazard problem; hiding information from the other party requires greater effort than before.

It is difficult to implement continuous monitoring and communication without digital means. There are numerous related digital solutions: mobile applications for communication, wearables to collect vital signs and automatic medicine dispensers to support adherence. Sensors have become smaller and cheaper and can be used to analyse sweat for example and monitor health indicators in this way [38]. A recent study also suggests that physicians increasingly utilize mobile applications to communicate with patients and that they also encourage the use of health applications [39].

Another related area is evidence-based medicine. It refers to making medication-related decisions systematically based on scientific evidence instead of relying on intuition [40]. Considering all available information and making decisions based on this reduces uncertainty and bias. The evidence-based approach can be practiced when specifying health policies as well as when treating individual patients. Digital tools are crucial since the approach requires searching and analysing large data sets (e.g. [41]).

5.5   The Availability of Information on Medical Care on the Internet

As described above the healthcare delivery market is pervaded by uncertainty and information asymmetry between health care experts and patients. The issue is particularly severe in emerging countries with undeveloped healthcare systems and associated regulatory frameworks. People have also become reluctant to unconditionally trust experts and seek to be empowered as patients [36]. This has led to the expansion of informal health markets and related new channels.

These new channels typically rely upon digital communication and can be both mobile applications and social media related. There are more than 100,000 health and medication related mobile applications available [42]. Broadly speaking healthcare information has become widely available and it is increasingly used by people looking for medical advice. A recent study suggests that 61 % of the American adult population using the internet goes online for health related information [43]. According to this study health related information is in fact one of the most searched for topics on the Internet.

One specific strand of healthcare information is that related to medication. In addition to online medical information such as pharmaceutical databases, services have emerged for acquiring medication online as well as increasing cost transparency and providing details on generic and alternative medicines (e.g. [30]). This kind of service has the potential to decrease information asymmetry in the medical market especially as this relates to pharmacies and their customers.

5.6   Medicine Demand Management

One topic that was raised in the interview conducted with a pharmaceutical industry representative [11] is that related to the medicine spoilage in hospital surroundings. This has severe economic consequences; according to the WHO's report the losses from inventory, including spoilage, can exceed 4 % of total medicine acquisition costs for instance [15].

The problem can derive from inefficient management practices [44]. Spoilage can result when medicines are ordered in excessive quantities or in non-optimal package sizes. Medicine shortages, on the other hand, can result in the use of expensive alternatives or in patients remaining hospitalized if medical treatment is not available.

Medicine demand management in hospitals has not been studied widely from an information systems perspective. Although the problem is not directly related to information asymmetry it is a consequence of a lack of information and results in transaction costs and economic inefficiency. It can be argued that spoilage can be reduced with better information management.

This requires combining relevant data from internal sources such as historical data on medicine consumption in hospitals along with data from external sources to help predict future consumption. The external sources could be those shared with other health care providers and the pharmaceutical industry. The information could be used to more accurately specify the quantities of medicines to be acquired. This process will be supported by appropriate digital tools.

One related area is the forecast of vaccine demand. For example, Finland faced a shortage of influenza vaccines in 2015, whereas the previous year the spoilage was 200,000 influenza shots [45]. The prediction models are typically based on age and similar coarse demographic variables [46]. These however do not take into account the public attitude towards vaccination.

Whereas the public sentiment can be an important factor when predicting the demand, it cannot be derived from a typical internal history database. Instead, this information could be extracted from external sources such as social and other media. This in fact has already been tested in United Nations Global Pulse project, in which the public immunization and anti-immunization sentiments were successfully tracked on the basis of social media debates [47].

6 Conclusion

In this article we have examined the digitalization of medical market through the concept of value network. We have also introduced some digital innovations that can be utilized to manage information asymmetries in the value network. It can be argued that equalizing information supply between actors in a value network is one of the benefits achieved through digitalization. It is described below how the discussed innovations map to different information asymmetries introduced in Sect. 4.

- Physician – patient: more information available in Internet, outcomes-based approach.
- Patient – physician: ePrescription, automatic medicine dispenser.
- Patient – insurance firm: outcomes-based approach.
- Patient – pharmacy: ePrescription.
- Pharmacy – patient: generic substitution.
- Service provider – medicine wholesalers: medicine demand mangement.

We can consequently argue that these solutions, which are digital or supported by digital means, can indeed be utilized to manage incomplete information and information asymmetries in the medical and health value network, and hence answer the research questions introduced in Sect. 1.

Electronic prescription has the potential to provide more complete information on patient medication history and hence provide more information to help identify increasingly efficient treatments. Generic substitution supported by digital technologies on the other hand will equalize information asymmetry relating to medicine costs. Value-based methods and increased health and medication information will help to equalize the highly unequal power relationship between health care experts and patients.

Digitalization will also reduce information asymmetry conversely; monitoring patients' compliance with prescriptions using digital tools will undermine patients' ability to disguise or withhold important information from health care experts. Lastly digitalization can enhance medicine demand management and hence reduce medicine spoilage and related inefficiencies.

The real economic impact of digitalization should however be more thoroughly studied in these cases. This is an exploratory study and obviously deeper research to investigate the area is required. Another obvious shortcoming of the article is to omit regulatory institutions, such as governmental and professional bodies, from this model. They have earlier been key actors in the management of asymmetric information. How they support the digitalized market is however not clear and requires research.

The healthcare market has not utilized digitalization as widely as many other lines of business. This can be concluded by comparing ICT investments in different sectors; ICT spend is much lower in relative terms in the health care sector than in sectors such as finance or manufacturing [2]. Undoubtedly EHR systems and such have been widely adopted but these only form the foundation for digitalization. In finance sector money has become virtual for more than a decade and customers are interfaced through digital channels. Similar far-reaching changes are yet to be seen in the medical market.

Costs continue to rise however and therefore there is a growing interest in finding solutions that increase efficiency. We have shown that digitalization could play a role in this and hence it should be one of the key areas of future research.

References

1. Brennen, S., Kreiss, D.: Digitalization and Digitization. Culture Digitally (2014). http://culturedigitally.org/2014/09/digitalization-and-digitization/
2. Pohjola, M.: Suomi uuteen nousuun. Teknologiateollisuu ry (2014)
3. Sherer, S.A.: From supply-chain management to value network advocacy: implications for e-supply chains. Supply Chain Manag. Int. J. **10**(2), 77–83 (2005)
4. Fjelstad, Ø.D., Ketels, C.H.M.: Competitive advantage and the value network configuration: making decisions at a Swedish life insurance company. Long Range Plan. **39**(2), 109–131 (2006)
5. Allee, V.: Value network analysis and value conversion of tangible and intangible assets. J. Intellect. Cap. **9**(1), 5–24 (2008)

6. Spence, M.: The Inexorable Logic of the Sharing Economy. Project-Syndicate, 28 September 2015. https://www.project-syndicate.org/commentary/inexorable-logic-sharing-economy-by-michael-spence-2015-09

7. Svorny, S.V.: Asymmetric Information and Medical Licensure. Cato Unbound. Cato Institute (2015). http://www.cato-unbound.org/2015/04/10/shirley-v-svorny/asymmetric-information-medical-licensure

8. Akerlof, G.A.: The market for 'lemons': quality uncertainty and the market mechanism. Q. J. Econ. **84**(3), 488–500 (1970)

9. Arrow, K.J.: Uncertainty and the welfare economics of medical care. Am. Econ. Rev. **53**(5), 941–973 (1963)

10. Miles, M.B., Huberman, M.A., Saldana, J.: Qualitative Data Analysis, 2nd edn. SAGE Publications, London (1994)

11. Merikallio, J.: Digitalization of the medical supply chain. [interv.] Peltoniemi, T., Suomi, R., 3 September 2015

12. Britnell, M.: Transforming Health Care Takes Continuity and Consistency. Harvard Business Review, 28 December 2015. https://hbr.org/2015/12/transforming-health-care-takes-continuity-and-consistency

13. Cebul, R.D., Rebitzer, J.B., Taylor, L.J., Votruba, M.: Organizational Fragmentation and Care Quality in the U.S Healthcare System. National Bureau of Economic Research, Cambridge (2008)

14. Burns, L.R., DeGraaff, R.A., Danzon, P.M., Kimberly, J.R., Kissick, W.L., Pauly, M.V.: The Wharton School Study of the Health Care Value Chain. The Wharton School, Philadelphia (2002)

15. World Health Organization: Analyzing and controlling pharmaceutical expenditures. WHO (2012)

16. eHealth Initiative: Electronic Prescribing: Toward Maximum Value and Rapid Adoption: Recommendations for Optimal Design and Implementation to Improve Care, Increase Efficiency and Reduce Costs in Ambulatory Care. eHealth Initiative, Washington D.C. (2004)

17. Abramson, E.L., Barrón, Y., Quaresimo, J., Kaushal, R.: Electronic prescribing within an electronic health record reduces ambulatory prescribing errors. Joint Comm. J. Qual. Patient Saf. **37**(10), 470–478(9) (2011)

18. Johnson, K.B., Lehmann, C.U.: Electronic prescribing in pediatrics: toward safer and more effective medication management. Off. J. Am. Acad. Pediatr. **131**(4), e1350–e1356 (2013)

19. Caldwell, N.A., Power, B.: The pros and cons of electronic prescribing. Arch. Dis. Child. **97**, 124–128 (2011)

20. Clyne, B., Bradley, M.C., Hughes, C., Fahey, T., Lapane, K.L.: Electronic prescribing and other forms of technology to reduce inappropriate medication use and polypharmacy in older people: a review of current evidence. Clin. Geriatr. Med. **28**(2), 301–322 (2012)

21. Lizano-Díez, I., Modamio, P., López-Calahorra, P., Lastra, C., Segú, J., Gilabert-Perramon, A., Mariño, E.: Evaluation of electronic prescription implementation in polymedicated users of Catalonia, Spain: a population-based longitudinal study. BMJ Open **4**(11), e006177 (2014). doi:10.1136/bmjopen-2014-006177

22. Kansallinen Terveysarkisto. www.omakanta.fi

23. Koskinen, J.: Datenherrschaft - an ethically justified solution to the problem of ownership of patient information. Turku School of Economics (2016)

24. Kuikkaniemi, K.: Ihmiskeskeinen vai yrityskeskeinen ratkaisu henkilökohtaisen datan hyödyntämiseen? Sytyke-lehti, vol. 1/2014 (2014)

25. Finlex: Laki lääkelain muuttamisesta 80/2003. Finlex (2003)

26. Kela: Lääkevaihto ja viitehintajärjestelmä. Kansaneläkelaitos (2015). http://www.kela.fi/laakevaihto-ja-viitehintajarjestelma
27. Koskinen, H., Martikainen, J.E., Maljanen, T., Ahola, E., Saastamoinen, L.K.: Viitehintajärjestelmän vaikutus järjestelmään kuuluvien ja sen ulkopuolella olevien lääkkeiden kustannuksiin. In: Klavus, J. (ed.) Terveystaloustiede, pp. 28–31. Terveyden ja hyvinvoinnin laitos (2013)
28. Fimea, K.: Finnish Statistics on Medicines. Edita Prima, Helsinki (2014)
29. Stenner, S.P., Chen, Q., Johnson, K.B.: Impact of generic substitution decision support on electronic prescribing behavior. J. Am. Med. Inform. Assoc. 17(6), 681–688 (2010)
30. myDrugCosts Inc. http://www.mydrugcosts.com
31. Martin, L.R., Williams, S.L., Haskard, K.B., DiMatteo, M.R.: The challenge of patient adherence. Ther. Clin. Risk Manag. 1(3), 189–199 (2005)
32. Foo, M.-H., Chua, J.C., Ng, J.: Enhancing medicine adherence though multifaceted personalized medicine management. In: 13th IEEE International Conference on e-Health Networking Applications and Services (Healthcom), pp. 262–265 (2011)
33. Gadkari, A.S.: Unintentional non-adherence to chronic prescription medications: how intentional is it really? BMC Health Serv. Res. 12, 98 (2012)
34. Suomi, R., Peltoniemi, T., Niinistö, J., Apell, M.: Safe home medicine delivery (Forthcoming)
35. Suomi, R., Li, H.: Innovation path in e-health - illustration in three Finnish medicine delivery initiatives. In: Proceedings of the WHICEB 2014, paper 69, pp. 672–679 (2014)
36. Bloom, G., Standing, H., Lloyd, R.: Markets, information asymmetry and health care: towards new social contracts. Soc. Sci. Med. 66(10), 2076–2087 (2008)
37. Gilmore, A.S., Zhao, Y., Kang, N., Ryskina, K.L., Legorreta, A.P., Taira, D.A., Chung, R.S.: Patient outcomes and evidence-based medicine in a preferred provider organization setting: a six-year evaluation of a physician pay-for-performance program. Health Serv. Res. 42(6), 2140–2159 (2007)
38. Howell, E.: US Military's Skin Sensors Could Reveal Science of Sweat. Live Science (2014). http://www.livescience.com/46072-military-skin-sensors.html
39. PR Newswire: Stethoscopes and Smartphones: Physicians Turn to Digital Tools to Boost Patient Outcomes - New Manhattan Research Study from Decision Resources Group Helps Marketers Understand How Physicians Are Using Digital Media and Technology in Their Practices. PR Newswire (2014). http://www.prnewswire.com/news-releases/stethoscopes-and-smartphones-physicians-turn-to-digital-tools-to-boost-patient-outcomes-261089461.html
40. Doherty, S.: Evidence-based medicine: arguments for and against. Emerg. Med. Australas. 17(4), 307–313 (2005)
41. Cochrane Library. http://www.cochranelibrary.com/
42. Lupton, D., Jutel, A.: 'It's like having a physician in your pocket!' A critical analysis of self-diagnosis smartphone apps. Soc. Sci. Med. 133, 128–135 (2015)
43. Lin, C.A., Atkin, D.J., Cappotto, C., Davis, C., Dean, J., Eisenbaum, J., House, K., Lange, R., Merceron, A., Metzger, J., Mitchum, A., Nicholls, H., Vidican, S.: Ethnicity, digital divides and uses of the Internet for health information. Comput. Hum. Behav. 51(Part A), 216–223 (2015)
44. Saedi, S., Kundakcioglu, E.O., Henry, A.C.: Mitigating the impact of drug shortages for a healthcare facility: an inventory management approach. Eur. J. Oper. Res. 251(1), 107–123 (2016)
45. Heikura, M.: Influenssarokotteet loppumassa kesken – THL ihmeissään: "Ei pitäisi olla mahdollista". YLE Uutiset (2015). http://yle.fi/uutiset/influenssarokotteet_loppumassa_kesken__thl_ihmeissaan_ei_pitaisi_olla_mahdollista/8485771

46. Keeling, M., Tildesley, M., House, T., Danon, L.: The Mathematics of vaccination. Math. Today **49**, 40–43 (2013)
47. United Nations Global Pulse: Understanding immunization awareness and sentiment through analysis of social media and news content. United Nations (2015)

Digital Health Literacy as Precondition for Sustainable and Equal Health Care – A Study Focussing the Users' Perspective

Anna-Lena Pohl[✉] and Roland Trill

Institute for eHealth and Management in HealthCare,
Flensburg University of Applied Sciences, Flensburg, Germany
{Anna-Lena.Pohl,Trill}@hs-flensburg.de

Abstract. Health Literacy and digital Health Literacy of citizens is a crucial topic in regarding equal access to health care. The empowerment of people ensures independent and self-reliable citizens who actively take part in health management. The use of eHealth services is one important tool in enabling people to develop the resources they need. In this respect it is crucial to put the needs, skills, cognitive capacities and personal use contexts of the end users in the focus of the consideration of eHealth services. Tailoring the services is one way to ensure sustainable use. As a result, the demand for smart and detailed measurement tools of eHealth Literacy is high, although the development is still in its infancy. Thus, beside the challenge of measuring and assessing the eHealth Literacy of people, the users' perspective is the most relevant angle in discussing eHealth services.

Keywords: eHealth Literacy · Health Literacy · Measurement tools · Acceptance · Equal access · Acceptance of eHealth

1 Introduction

In today's society electronic health services play an increasing role and generally in the western societies there is a great willingness to use them [1–3]. EHealth services are able to offer a variety of advantages to the user [4–6]. Eland-de Kok et al. found in a systematic review that eHealth interventions for chronically ill persons can lead to positive effects on primary health outcomes [7]. Santana and colleagues measured that almost 27 % of European citizens who had searched for health information online also have made active suggestions on diagnosis or treatment to their physician and thus took a more active role in medical decision making [6]. In a meta-analysis of randomized controlled trials on the effects of consumer health information technologies for diabetes patients, Or and Tao found that the usage of eHealth technologies reveal positive effects on clinical parameters such as blood pressure or cholesterol levels [8].

Nevertheless the literature shows that eHealth services often are not accepted by the intended users (e.g. Google Health) [9] at all or that the interest is flagging over time [10]. If health services on the Internet are not used properly it might lead to emotional harm of the user or, in one reported case to the death of a patient [11]. Numerous factors contributing to an appropriate use of eHealth services include different facets:

© Springer International Publishing Switzerland 2016
H. Li et al. (Eds.): WIS 2016, CCIS 636, pp. 37–46, 2016.
DOI: 10.1007/978-3-319-44672-1_4

There might be different contexts of use (e.g. support from other persons) [12], different personalities (e.g. high intrinsic motivation, anxiety) [12, 13], and characteristics of the intended users (e.g. gender, age) [12] or diverse competencies of the users to use eHealth services [14, 15].

Also for eHealth system designer in-depth knowledge about needs, capabilities and contexts of use of the consumers is essential to improve effectiveness and applicability of eHealth services. This points explicitly at the level of eHealth literacy of citizens. Investigating this issue further contributes to better-targeted, equitable and safe eHealth tools maximizing empowerment and health outcomes for citizens while reducing health inequalities [16].

Due to ubiquitous accessible health information and interactive functions eHealth services are expected to help overcome unequal access to health care and thus help decrease social inequalities in health care. Nevertheless, we face the risk that individuals will not use them in the most efficient way simply because they are not able to. Thus, it is essential to understand what skills are needed to use eHealth services efficiently.

It would be desirable if the discussion about the proven coherence between these factors could raise awareness of eHealth literacy issues among policymakers involved in developing strategies for implementation of eHealth services.

2   Challenges in Health Care 2016–2025

In the WHO European Region the percentage of the population aged 65 years of age or older is rising, a trend that is likely to continue. This will increase the demand of health services that can handle multiple disorders with functional, psychological, and social dimensions. The specific WHO objective for health services in Europe is that "people in the region should have much better access to family- and community-oriented primary health care, supported by a flexible and responsive hospital system" [17]. These services should further be based on multi-professional teams that also includes informal carers, and that ensure the individual's active participation.

At the same time the pressure on health care budgets increases and all countries face the situation that they need to balance an increasing demand for health care services and at the same time a decreasing supply.

This situation is even tougher in rural areas where distances are much farther and less GPs are available.

For some people these factors alone lead to a limited access to health care. In such a situation the competences of each person are even more important to be able to search and find information on their own, mostly written sources in the world wide web.

3   Health Literacy – A Predictor for Better Health and Behaviour Changes

Health Literacy is a term that was first introduced over 30 years ago [18]. Ratzan and Parker created the mostly used definition; according to them Health Literacy is "the degree to which individuals have the capacity to obtain, process, and understand basic health information and services needed to make appropriate health decisions" [19].

Since the 1970s the concept of Health Literacy has been used widely in research and can be measured by a large variety of tools (e.g. TOFHLA, HALS, REALM, MART, FHLM, ELF...) [20]. One problem with these tools is that although they offer gold-standard for the measurement of Health Literacy (like TOFHLA and REALM), it is not possible to apply them for computer-based use [21].

In his article "Health Literacy as a public health goal" Nutbeam defined three levels of health literacy [22]:

Basic/Functional Literacy: Skills that allow an individual to read consent forms, medicine labels, and health care information and to understand written and oral information given by physicians, nurses, pharmacists, or other health care professionals and to act on directions by taking medication correctly, adhering to self-care at home, and keeping appointment schedules.

Communicative/Interactive Literacy: The wide range of skills, and competencies that people develop over their lifetimes to seek out, comprehend, evaluate, and use health information and concepts to make informed choices, reduce health risks, and increase quality of life.

Critical Literacy: Strengthening active citizenship for health by bringing together a commitment to citizenship with health promotion and prevention efforts and involving individuals in: understanding their rights as patients and their ability to navigate through the health care system; acting as an informed consumers about the health risks of products and services and about options in health care providers, and acting individually or collectively to improve health through the political system through voting, advocacy or membership of social movements. This refers to an encompassing empowerment of people.

These competencies enable citizens to act more self-determined and to make choices in their own health care. Thus, health literacy has a great practical value in achieving better health outcomes among those who actively participate in healthcare decisions compared to those who do not. Additionally, research on health literacy has shown that a poor health literacy is closely related to a reduced responsiveness to health education and use of disease prevention services, a decreased ability to manage chronic diseases and to an increase of healthcare costs [23, 24]. On the other hand a low socio-economic status has been identified as a risk factor for a low level of health literacy [24]. Even on European Union level the impact of inadequate health literacy on health and wellbeing and the correlation with inequities in health issues is meanwhile a high ranked topic on the agenda. The *Europe 2020 Strategy* state that reducing poverty, the risk of poverty and social exclusion is one of the five main targets [26]. The above mentioned social indicators are seen as important factors associated with inadequate levels of health literacy.

Since the number of mobile devices in use exceeds the number of people in many countries the role of the internet in health information seeking will increase steadily. Thus, the digital skills in general and the eHealth literacy of citizens need to be carefully addressed [25].

4   eHealth Services and eHealth Literacy – State of the Art

Due to this background an extension of the Health Literacy concept to include e-health related competencies was performed. Norman and Skinner were pioneering in the concept of eHealth Literacy: They defined it as "the ability to seek, find, understand, appraise health information from electronic sources and apply the gained knowledge to addressing or solving a health problem" [27].

Whereas Health Literacy measures competencies in the context of paper-based resources in the healthcare environment, eHealth Literacy is much more complex: Persons who are intended to use electronic sources for health purposes need a variety of skills – basic literacy (reading and writing of texts) is as well necessary as knowing how to use computers and understand and evaluate science and media [27]. So Norman and Skinner defined eHealth Literacy not just as a combination out of the capability to use computers and Health Literacy but as a meta-literacy out of different facets of literacy.

Thus, eHealth Literacy consists of six different literacies:

- Health literacy: Health knowledge comprehension
- Computer literacy: Skills to use hard- and software to solve problems
- Science literacy: Understanding science processes and outcomes
- Traditional literacy: Reading, writing, and numeracy, which is important as electronic sources of health information are still text dominant.
- Media literacy: Thinking critically about media content
- Information literacy: Seeking and understanding information to make decisions

The competencies are grouped in the Lily model which is shown in Fig. 1. They are not stable but might increase over time [27] thus enabling the training of said literacy [28].

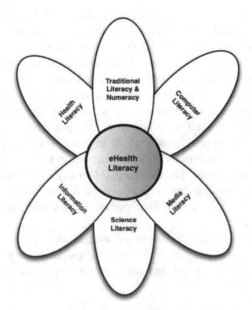

Fig. 1. eHealth literacy: the Lily Model

Norman who, together with Skinner, created the Lily Model in 2006 wrote five years later that he sees some problems with it, for example that it does not fit to the Web 2.0 solutions [29]. Others stated that eHealth literacy was heavily depending on social structures [30] or the individual motivation to use a system [13] which is not included in the original Lily Model. Thus, the original Lily Model is lacking several aspects of eHealth usage such as the contexts of use, further user characteristics like anxiety or motivation, and different personalities of the intended users including age, gender and socioeconomic status.

In her article "Toward a Comprehensive Model of eHealth Literacy" for the PAHI workshop 2014 [31] Gilstad describes how she included other literacies like cultural, contextual, and communicative competencies into the Lily Model. Furthermore, she emphasized that it is important to distinguish between knowledge generated from theoretical sources such as books or research articles and procedural knowledge gained from practical experiences.

In their recently published paper Kayser et al. [16] presented a new concept of eHealth Literacy which consists of three overarching themes considering seven domains related to the end-user capabilities, access to technology and experiences in using technology.

Currently, the level of Health Literacy and eHealth Literacy of vulnerable groups, e.g. migrants is in the focus. According to Sørensen et al. health literacy is closely linked to the level of education and competences and motivation to search for information, to understand and use them in e.g. therapy decision [32, 33]. This also applies to decisions regarding the participation in prevention and health promotion measures and thus has a significant impact on life quality and sustainable health outcomes. International studies show that the health literacy of vulnerable groups (older people, migrants, people with low socio-economic status) is lower than in other groups [34]. To tackle this challenge customized and targeted interventions are needed to address especially these vulnerable groups.

5   Measuring and Evaluating eHealth Literacy

Since 2010 in Germany the ePatient Survey [35] collects data about health related internet usage of citizens. Patients, chronically ill persons and people surfing for health related information in the world wide web are interviewed regarding their internet use and the effect on their health behaviour and treatment adherence.

Main findings of the ePatient Survey 2015 were:

- The use of digital health services by citizens and patients has positive effects on coping with diseases, treatment and medication adherence, discussions with health professionals, understanding of behavioural rules and the cooperation with other actors such as health insurances or rehabilitation facilities.
- Because of the parallel existence of the traditional health care system and the digital health care market citizens/patients would like to get recommendations for digital services from their GPs or health insurances.

- The comprehensibility, the usability and the readability of digital services are the biggest concerns of the users. Thus, digital health services often lose impact and intervention potential because users deny these attributes.

The just published report from 2016 shows that not age but the level of education is crucial for online health information seeking. Differently than possibly expected the average age of health information seekers is 59 and thus higher than the demographic average. 54 % of all people seeking for health information online are female. 32 % of the interviewees have a university degree and 35 % a secondary school certificate. 43 % are chronically ill, only 29 % are seeking for health information in an acute case [35].

This years' survey shows, that people seeking health information online do not only search for information but use these informations for their therapy decision. Furthermore, the usage of apps has an effect on therapy and medication adherence. 46 % of those questioned state that with the help of a medication app they can significantly better manage their medication intake, while 30 % state that they can deal slightly better with own medication management.

The willingness to pay for online health information services or apps increased since 2015. While last year 80 % were not willing to pay this number decreased in 2016 and now 50 % of the user state that they are not willing to pay for such services [35].

Thus, the survey gives a good overview about user habits and effects of internet usage on health outcomes. By measuring an individual's eHealth Literacy it is possible to get an overview of his or her competencies. Furthermore, the functionality of e-health application software can be evaluated by measuring the changing of competencies over the time this software is used.

Until now eHEALS (eHealth Literacy Scale), [36] is the widest used measurement tool for assessing e-health literacy of individuals [21], consisting of 8–10 items. Norman and Skinner developed it in English using a sample of Canadian adolescents [14, 36]. Koo et al., van der Vaart et al., and Mitsutake et al. translated eHEALS into Chinese, Dutch, respectively Japanese [30, 37, 38]. Soellner et al. provided a German translation of the eHEALS [39].

Also Hove et al., Ghaddar et al., and Paek and Hove used eHEALS to measure eHealth Literacy of adolescents [40–42]. Neter et al. reduced the number of items to only six to measure eHealth Literacy in the average Israeli adult population [14].

Another broad approach of eHEALS to measure eHealth Literacy of a larger group of people is the work of Mitsutake et al. who measured an association of approximately 3000 Japanese adults with their knowledge about colorectal cancer [43].

Furthermore Ossebaard et al. measured the eHealth related literacy for patients with chronic diseases. This study is one of the few found papers that used eHEALS outside of the North American area (Netherlands) [44].

Also eHealth Literacy of low-income parents with chronically ill children or with children that are in a pediatric palliative care program, HIV patients and older adults has been measured using eHEALS [45–49].

Tennant and Stellefson found that baby boomers and older persons who used Health 2.0 technologies had higher levels of eHealth Literacy than persons who did not. For their study they as well used the eHEALS [50].

An alternative measurement of eHealth literacy was proposed by Chan et al. who developed a taxonomy to characterize the complexity of several eHealth tasks and therefore draw conclusions on the individual users' competencies to perform those tasks [51]. This approach nevertheless is very complex and time consuming as it implies the direct observation of individual persons during their usage of an eHealth system. In our approach including three spatially separated user sites with a large number of users it was not practicable to perform such an observational study. In 2014, Chew published a conference paper dealing with the development of a new scale to measure eHealth literacy [52].

In 2013 Jones developed the Patient eHealth Readiness Scale (PERQ) which includes items from the eHEALS as well as contextual factors like Internet use, support from other persons and demographics such as age and gender [53]. This approach has been used two times in published papers [54, 55].

The work of Chew has not been tested in the practice yet thus does not deliver starting points for its usage. To adequately address the finding that the eHealth Literacy concept does not include all relevant factors explaining the use of interactive eHealth solutions a broader range of factors should be included in a measurement tool to adequately assess eHealth Literacy.

There has been done some international research in measuring eHealth Literacy but still all tools lack the acknowledgement of different personal backgrounds that influence deeply the measured competencies: social and cultural factors need to be taken into account when discussing the level of eHealth Literacy.

6 Conclusion

There is an urgent need to revise the understanding of eHealth literacy and its measurement tools. Since current analysis lacks several important aspects regarding further eHealth barriers such as personality factors like anxiety or trust the used measurement tools do not meet requirements of the technological development of the internet and social media. Additionally, cultural and socio-economic aspects have to be taken into account.

The above outlined relationship between literacy and health should lead to an increase of patient participation in health care. Many countries took up the slogan of an actively participating and empowered patient. This aim will remain an illusion as long as policy will not invest more financial and personal resources into smart interventions improving the eHealth literacy of all population groups. Otherwise they will only reach a subset of citizens, mostly those who already have a fair level of eHealth literacy and access to prevention and health promotion measures. And even worse, without specific attention on vulnerable groups and those with low eHealth literacy the way might lead to an increase of health inequalities.

Future research and projects should focus on developing and testing reliable measurement tools for eHealth Literacy. Based on such an assessment of the current level of eHealth Literacy it is necessary to place greater emphasis on user-centred interventions to raise the level of the eHealth Literacy. Specific needs and requirements of the target group should be in the focus of such intervention planning.

References

1. Andreassen, H., Bujnowska-Fedak, M., Chronaki, C., Dumitru, R., Pudule, I., Santana, S., Voss, H., Wynn, R.: European citizens' use of E-health services: a study of seven countries. BMC Public Health **7**, 53 (2007)
2. http://www.fiercemobilehealthcare.com/story/physicians-split-use-mhealth-apps/2014-02-24?utm_medium=nl&utm_source=internal
3. http://www.pewinternet.org/2013/01/15/health-online-2013/
4. Eysenbach, G.: What is e-health? J. Med. Internet Res. **3**, e20 (2001)
5. Holmström, I., Röing, M.: The relation between patient-centeredness and patient empowerment: a discussion on concepts. Patient Educ. Couns. **79**, 167–172 (2010)
6. Santana, S., Lausen, B., Bujnowska-Fedak, M., Chronaki, C., Prokosch, H., Wynn, R.: Informed citizen and empowered citizen in health: results from an European survey. BMC Family Pract. **12**, 20 (2011)
7. Eland-de Kok, P., van Os-Medendorp, H., Vergouwe-Meijer, A., Bruijnzeel-Koomen, C., Ros, W.: A systematic review of the effects of e-health on chronically ill patients. J. Clin. Nurs. **20**, 2997–3010 (2011)
8. Or, C.K., Tao, D.: Does the use of consumer health information technology improve outcomes in the patient self-management of diabetes? A meta-analysis and narrative review of randomized controlled trials. Int. J. Med. Inform. **83**, 320–329 (2014)
9. http://ddormer.wordpress.com/2011/06/29/lessons-from-google-health/, http://www.himss.org/News/NewsDetail.aspx?ItemNumber=4425
10. Kelders, M.S., Van Gemert-Pijnen, E.W.C.J., Werkman, A., Nijland, N., Seydel, R.E.: Effectiveness of a web-based intervention aimed at healthy dietary and physical activity behavior: a randomized controlled trial about users and usage. J. Med. Internet Res. **13**, e32 (2011)
11. Crocco, A.G., Villasis-Keever, M., Jadad, A.R.: Analysis of cases of harm associated with use of health information on the internet. JAMA **287**, 2869–2871 (2002)
12. Venkatesh, V., Morris, M.G., Gordon, B.D., Davis, F.D.: User acceptance of information technology: toward a unified view. MIS Q. **27**, 425–478 (2003)
13. Bodie, G.D., Dutta, M.J.: Understanding health literacy for strategic health marketing: eHealth literacy, health disparities, and the digital divide. Health Mark. Q. **25**, 175–203 (2008)
14. Neter, E., Brainin, E.: eHealth literacy: extending the digital divide to the realm of health information. J. Med. Internet Res. **14**, e19 (2012)
15. van Dijk, J.A.G.M.: The Deepening Divide: Inequality in the Information Society. SAGE Publications, Beverly Hills (2005)
16. Kayser, L., Kushniruk, A., Osborne, R., Norgaard, O., Turner, P.: Enhancing the effectiveness of consumer-focused health information technology systems through ehealth literacy: a framework for understanding users' needs. J. Med. Internet Res. Hum. Fact. **2**, e9 (2015)
17. Gröne, O., Garcia-Barbero, M.: Integrated care: a position paper of the WHO European office for integrated health care services. Int. J. Integr. Care **1**, e21 (2001)
18. Simonds, S.K.: Health education as social policy. Health Educ. Monogr. **2**, 1–25 (1974)
19. Ratzan, S.C., Parker, R.: Introduction. In: Seldon, C., Zorn, M., Ratzan, S.C., Parker, R. (eds.) National Library of Medicine Current Bibliographies in Medicine: Health Literacy. National Institutes of Health, US Department of Health and Human Services, Washington, DC (2000)
20. http://www.nchealthliteracy.org/instruments.html

21. Collins, S.A., Currie, L.M., Bakken, S., Vawdrey, D.K., Stone, P.W.: Health literacy screening instruments for eHealth applications: a systematic review. J. Biomed. Inform. **45**, 598–607 (2012)
22. Nutbeam, D.: Health literacy as a public health goal: a challenge for contemporary health education and communication strategies into the 21st century. Health Promot. Int. **15**, 259–267 (2000)
23. Protheroe, J., Nutbeam, D., Rowlands, G.: Health literacy: a necessity for increasing participation in health care. Br. J. Gen. Pract. **59**, 721–723 (2009)
24. Dewalt, D.A., Berman, N.D., Sheridan, S., et al.: Literacy and health outcomes: a systematic review of the literature. J. Gen. Intern. Med. **19**, 1228–1239 (2004)
25. Quaglio, G., Sørensen, K., Rübig, P., Bertinato, L., Brand, H., Karapigeris, T., Dinca, I.: Accelerating the health literacy agenda in Europe. Health Promot. Int. (2016). doi:10.1093/heapro/daw028
26. http://ec.europa.eu/europe2020/europe-2020-in-a-nutshell/index_en.htm
27. Norman, C.D., Skinner, H.A.: eHealth literacy: essential skills for consumer health in a networked world. J. Med. Internet Res. **8**, e9 (2006)
28. Xie, B.: Improving older adults' e-health literacy through computer training using NIH online resources. Libr. Inf. Sci. Res. **34**, 63–71 (2012)
29. Norman, C.: eHealth literacy 2.0: problems and opportunities with an evolving concept. J. Med. Internet Res. **13**, e125 (2011)
30. van der Vaart, R., van Deursen, A.J., Drossaert, C.H., Taal, E., van Dijk, J.A., van de Laar, M.A.: Does the eHealth literacy scale (eHEALS) measure what it intends to measure? Validation of a Dutch version of the eHEALS in two adult populations. J. Med. Internet Res. **13**, e86 (2011)
31. Gilstad, H.: Toward a comprehensive model of eHealth literacy. In: 2nd European Workshop on Practical Aspects of Health Informatics, pp. 63–72 (2014)
32. Sørensen, K., Van den Broucke, S., Doyle, G., Pelikan, J., Slonska, Z., Brand, H.: Health literacy and public health. A systematic review and integration of definitions and models. BMC Public Health **12**, 80 (2012)
33. Quenzel, G., Schaeffer, D.: Health Literacy – Gesundheitskompetenz vulnerabler Bevölkerungsgruppen. Bielefeld (2016)
34. Sørensen, K., Pelikan, J.M., Röthlin, F., Ganahl, K., Slonska, Z., Doyle, G., Fullam, J., Kondilis, B., Agrafiotis, D., Uiters, E., Falcon, M., Mensing, M., Tchamov, K., Van den Broucke, S.V., Brand, H.: HLS-EU consortium: health literacy in Europe: comparative results of the European health literacy survey (HLS-EU). Eur. J. Public Health Adv. Access **25**, 1053–1058 (2015)
35. http://epatient-rsd.com/epatient-survey/
36. Norman, C.D., Skinner, H.A.: eHEALS: the eHealth literacy scale. J. Med. Internet Res. **8**, e27 (2006)
37. Koo, M., Norman, C.D., Chang, H.-M.: Psychometric evaluation of a Chinese version of the eHealth literacy scale (eHEALS) in school age children. Int. Electron. J. Health Educ. **15**, 29–36 (2012)
38. Mitsutake, S., Shibata, A., Ishii, K., Okazaki, K., Oka, K.: Developing Japanese version of the eHealth literacy scale (eHEALS)]. [Nihon koshu eisei zasshi. Jpn. J. Public Health **58**, 361–371 (2011)
39. Soellner, R., Huber, S., Reder, M.: The concept of eHealth literacy and its measurement: German translation of the eHEALS. J. Media Psychol. **26**, 29–38 (2014)
40. Hove, T., Paek, H.-J., Isaacson, T.: Using adolescent eHealth literacy to weight trust in commercial web sites: the more children know, the tougher they are to persuade. J. Advert. Res. **51**, 524–537 (2011)

41. Ghaddar, S.F., Valerio, M.A., Garcia, C.M., Hansen, L.: Adolescent health literacy: the importance of credible sources for online health information. J. Sch. Health **82**, 28–36 (2012)
42. Paek, H.J., Hove, T.: Social cognitive factors and perceived social influences that improve adolescent eHealth literacy. Health Commun. **27**, 727–737 (2012)
43. Mitsutake, S., Shibata, A., Ishii, K., Oka, K.: Association of eHealth literacy with colorectal cancer knowledge and screening practice among internet users in Japan. J. Med. Internet Res. **14**, e153 (2012)
44. Ossebaard, H.C., Seydel, E.R., van Gemert-Pijnen, L.: Online usability and patients with long-term conditions: a mixed-methods approach. Int. J. Med. Inform. **81**, 374–387 (2012)
45. Robinson, C., Graham, J.: Perceived Internet health literacy of HIV-positive people through the provision of a computer and Internet health education intervention. Health Inf. Libr. J. **27**, 295–303 (2010)
46. Knapp, C., Madden, V., Wang, H., Sloyer, P., Shenkman, E.: Internet use and eHealth literacy of low-income parents whose children have special health care needs. J. Med. Internet Res. **13**, e75 (2011)
47. Xie, B.: Effects of an eHealth literacy intervention for older adults. J. Med. Internet Res. **13**, e90 (2011)
48. Xie, B.: Experimenting on the impact of learning methods and information presentation channels on older adults' e-Health literacy. J. Am. Soc. Inf. Sci. Technol. **62**, 1797–1807 (2011)
49. Knapp, C., Madden, V., Marcu, M., Wang, H., Curtis, C., Sloyer, P., Shenkman, E.: Information seeking behaviors of parents whose children have life-threatening illnesses. Pediatr. Blood Cancer **56**, 805–811 (2011)
50. Tennant, B., Stellefson, M.: eHealth literacy and web 2.0 health information seeking behaviors among baby boomers and older adults. J. Med. Internet Res. **17**, e70 (2015)
51. Chan, C.V., Kaufman, D.R.: A framework for characterizing eHealth literacy demands and barriers. J. Med. Internet Res. **13**, e94 (2011)
52. Chew, F.: Developing a new scale to measure e-health literacy. In: Medicine 2.0 World Congress on Social Media, Mobile Apps, Internet/Web 2.0 (2014)
53. Jones, R.: Development of a questionnaire and cross-sectional survey of patient eHealth readiness and eHealth inequalities. Medicine 2.0 **2**, e9 (2013)
54. Jones, R.B., Ashurst, E.J., Atkey, J., Duffy, B.: Older people going online: its value and before-after evaluation of volunteer support. J. Med. Internet Res. **17**, e122 (2015)
55. LeRouge, C., Van Slyke, C., Seale, D., Wright, K.: Baby boomers' adoption of consumer health technologies: survey on readiness and barriers. J. Med. Internet Res. **16**, e200 (2014)

Factors Affecting the Availability of Electronic Patient Records for Secondary Purposes – A Case Study

Antti Vikström[1(✉)], Sanaz Rahimi Moosavi[1], Hans Moen[1], Tapio Salakoski[1], and Sanna Salanterä[2]

[1] Department of Information Technology, University of Turku, Turku, Finland
{anelvi,saramo,hanmoe,sala}@utu.fi
[2] Department of Nursing Science, University of Turku, Turku, Finland
sansala@utu.fi

Abstract. The purpose of this paper is to explore secondary use of Finnish electronic patient record (EPR) data in the context of clinical research and product development. Further, EPR availability enhancing procedures and technologies are analysed. The sensitive nature of patient data restricts the use and availability of EPR data in secondary purposes. A case study of secondary users of EPR data was conducted in Southwest Finland. Semi-structured interviews were used to evaluate the effectiveness of procedures and technologies implemented to protect EPR data. In total, 9 experts were interviewed from the fields of academic research, product development, and health management. The results show that three main factors affecting the availability of EPR data in secondary use are data management, privacy preserving, and secondary users. Challenges included in data management concerned the effect of demanding data request procedures and external information system service providers. Two privacy preserving approaches were identified: the use of altered data and protected EPR processing environment. These approaches provide higher availability or more valuable content, both affecting possible secondary users and use cases.

Keywords: Privacy preserving · Patient records · Secondary use · Open data

1 Introduction

The use of patient records is restricted in academic research and product development due to the sensitive nature of such records. Secondary use of patient records depends on accessing and processing confidential data initially not intended for that purpose. For such a secondary use, preserving the privacy and security is essential. The further use of patient records can be intensified to correspond with the actual needs by enhancing the data availability. Access to authentic health data enables new clinical research possibilities and offers valuable information to

H. Li et al. (Eds.): WIS 2016, CCIS 636, pp. 47–56, 2016.
DOI: 10.1007/978-3-319-44672-1_5

product development in health sectors. Successful secondary use adds value to the original data through, e.g., knowledge extraction and the development of tailored tools. This paper discusses availability enhancing procedures and technologies for secondary use of electronic patient record (EPR) data. Research considering the secondary use of personal records is limitedly available especially in Finland.

An EPR consists of structured data and unstructured narrative text. Structured data is typically coded in a systematic manner in order to enhance use efficiency [1]. Clinical text is mainly unstructured narrative text written by a clinician for the purpose of managing and summarizing patient care. Secondary use of unstructured narrative text is challenging since the information therein is expressed more freely when compared to structured data [2].

Patients' privacy is an essential part of healthcare ethics and therefore an important concept. Privacy refers to a fundamental right to keep confidential information private. There are several definitions of the concept and dimensions of privacy in healthcare context. The concept definition [3] used in this paper divides privacy into four dimensions which are physical, psychological, social, and informational. Physical privacy ensures individuals' right to control the extent a person is physically accessible to others. Psychological privacy stands for the right to think and form personal attitudes and values. Social privacy ensures person's ability to control social contacts. Informational privacy refers to person's right to decide what personal information can be shared to other groups, therefore strongly connecting to the topic of this study.

The data content and structure are quality factors that determine the value of an EPR in secondary use. In the context of this paper, privacy preserving refers to actions ensuring patients' privacy in secondary use of unstructured EPR data. Further, de-identification and anonymization are considered as privacy preserving measures. De-identification refers to deleting, masking, suppressing, or generalizing explicit identifiers within a data set [4]. Anonymization refers to actions for ensuring that data is not identifiable within the data set. Not identifiable content should not be linkable back to the specific entity [5]. Different privacy preserving measures may radically affect the usability of the data while ensuring data confidentiality.

There is no unambiguous definition for patients' rights in healthcare context. The concept is a combination of *legislation, principles,* and *professional guidelines.* Legislation states civil rights for patients and obligations for professional healthcare personnel. The foundation for patients' rights is established with universal agreements, followed principles, and reached consensus [3].

One widely used definition of the concept of information security refers to actions for maintaining confidentiality, integrity, and availability of the data [6]. The extensive use of modern information systems generates an enormous need for security solutions. Healthcare information systems containing sensitive data are a valid example of this development. Securing and protecting information systems against internal and external threats is essential when considering the concept of information security in healthcare domain [7]. A comprehensive security system should include at least five perspectives: *people, procedures, technology, physical environment,* and *legislation* [8].

2 Related Work and Motivation

There is only a limited number of studies concerning the secondary use of EPR data. In Finland, government initiatives have been launched to clarify eligible secondary users and use procedures of patient records. The Ministry of Social affairs and Health of Finland is preparing legislation regarding the secondary use of patient data during the years 2015 and 2016 [9]. The purpose of this legislation is to enable secondary use of patient data in research, management, and development or monitoring of operations.

"Open Science and Research Initiative" was launched by The Ministry of Education and Culture of Finland in 2014 [10]. The initiative promotes open science and information availability in research and has been developing a reference architecture for research regarding sensitive information. The purpose is to provide a foundation for specified architectures or infrastructures for handling sensitive information by defining essential users, principles, and procedures. Detailed descriptions or technical solutions are not considered in the reference architecture.

Legitimate secondary users for personal data are defined in the Data protection directive of the European Union (EU) [11]. The directive states that personal data may only be collected for specific, explicit, and legitimate purposes if the purpose of processing is compatible with the data collection. Data processing is also enabled for historical, statistical, and scientific purposes if the process is secured. In Finnish legislation, the data processing restriction is implemented in Personal data act, which allows personal data and identity number processing in historical, scientific research, and statistics [12]. Although the secondary use of personal data is highly restricted, the utilization of EPR data also arouses interest in users not specified in Finnish legislation including product developers and open access users.

In the United States of America (USA), Health Insurance Portability and Accountability Act of 1996 (HIPAA) defines that if personal health information (PHI) is removed or masked from EPR data in order to use the data in secondary purposes, patient's informed consent is not needed. "Safe Harbor" technique of HIPAA legislation states 18 types of PHI which should be protected [13]. HIPAA legislation does not explicitly cover all possible user entities, therefore leaving PHIs unprotected in some secondary use cases. Due to the lack of consistent policies and standards professional groups have been developing a national framework for secondary use of patient data in USA [14].

As presented above, the lack of clear procedures and technological approaches in literature motivates to carry out a study considering secondary use of EPR data in Finland. In this paper, data providers' and potential users' perspectives are used to explore the clarity and effectiveness of EPR data related procedures and technological approaches providing privacy preserving.

3   Method

This research was carried out as a case study of secondary users of EPR data in Southwest Finland. The Ikitik [15] consortium and its stakeholders were used for analysis. Ikitik consortium develops research-based information and language technology solutions for health information and communication. The consortium stakeholders include an EPR data provider and business partners.

A qualitative approach was used for data collection and analysis. Semi-structured interviews were used for collecting expert views in the context of this domain. The interviewees were selected based on their expertise in handling EPR data in academic research, product development, or health management. The motivation for these interviews was to evaluate the effectiveness of procedures and technology solutions used to protect EPR data, further resolving attitudes towards the enhancement of EPR availability in secondary purposes. Interviews were divided into two sections. Section 1 consists of questions and conversation topics concerning availability of EPR data. Section 2 considers privacy preserving measures in secondary use of EPR data. In total, 9 experts were selected for interviews including professors, physicians, and chief officers in health and business.

1. Professor, Nursing Science
2. Professor, Information Technology
3. Chief Medical Information Officer
4. Chief Physician of Research
5. Chief Information Officer
6. Chief Archivist and Data Protection Officer
7. Nurse Director
8. Branch Manager, Health IT-services
9. Chief Executive Officer, Language Technology

The data collection for this study was performed with in-person interviews using semi-structured questions and conversation topics. The audio of each interview was recorded with a mobile audio recording device for later analysis. To enhance the anonymity of the study, the interviewees are not cited directly in this paper. The interviews took place from December 2015 to March 2016.

The content analysis process [16] used in this paper utilizes both inductive and deductive approaches (see Table 1). During the deductive phase, analysis matrix was created using *technology* and *procedures* perspectives and the corresponding data was gathered. Following inductive phase involved data grouping, categorization, and abstraction.

4   Results and Discussion

Secondary Users. Motivation to allow open access to privacy preserved EPR data in secondary purposes was discussed with interviewees attending this study. Open data enables the development of new research approaches by providing

Table 1. Content analysis process

Procedures	Technology
"The data request procedure is extensive and unstructured."[1]	"The technical implementation for data delivery is outsourced."[2]

Group	Sub-category	Main category
1	Data request procedure	Data management
2	Data delivery procedure	Data management

data to purposes which are initially undefined. When considering formal sciences, open data is preferred over highly specified medical research basis. Open access to research data enhances the reliability of research results by making the use of fabricated results extremely difficult. Further use of EPR data in publicly funded healthcare also enhances the use efficiency in terms of common good. The overview of open data and product development related results is presented in Fig. 1 under "Secondary users" section, which presents potential users and use intentions in relation to identified use approaches.

Data providers have recognized the possibility of providing EPR data to open access use, but implementation procedures have not been defined. The lack of national level instructions enforces EPR data providers to progress conservatively in open data projects. Currently, open data use is only possible with coarse data sets, which affects radically the value of unstructured data. Still, coarse open data is being successfully used in pharmaceutical research. Open access to aggregated patient data allows pharmaceutical companies to investigate possible patient groups for research.

The motivation to use EPR data in product development was discussed with the data provider. Successfully operated product development process enhances products' cost-efficiency, therefore providing data owners a better product with better price. Product development is considered as a possible mode of secondary use and no data request is automatically rejected by the data provider. However, the representatives of product development do not typically consider EPR data to be available at all or only to a limited extent. Legislation and current procedures do not encourage business users to process actual EPR data in order to support business.

The purpose of using real unstructured EPR data in product development was criticised by the data owner, since artificial patient scenarios are available for business users. Therefore, secondary users in business would receive highly anonymized and mixed patient data to ensure the privacy of individual patients.

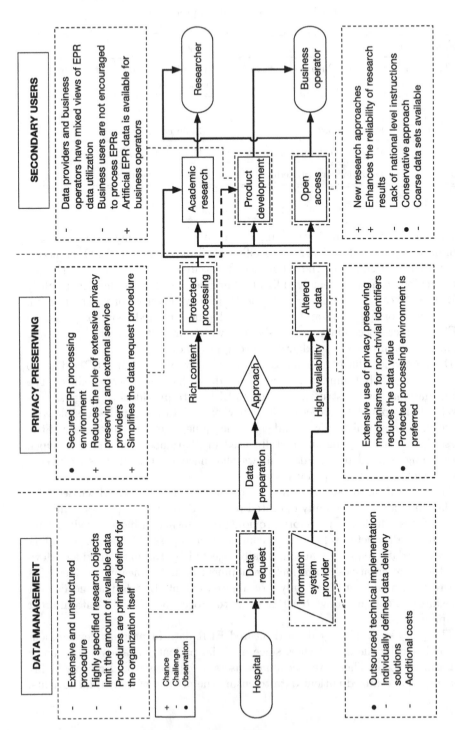

Fig. 1. Overview of study results

Data Management. The data request procedure is a crucial part of secondary use of EPR data. A clarified request procedure serves both the provider and potential users in terms of research intentions and technical solutions. Based on the discussions considering EPR data management, major topics in this context were identified: data request procedure, technological procedures, data storage, and data ownership (see Fig. 1 "Data management").

The interviewees representing academic research shared a critical view on data request procedure. Current request procedure is extensive and unstructured while demanding significant amount of labour. General level instructions for requests exist, but achieving these directives can be problematic due to interpretation issues. Although request instructions are on general level, research purposes need to be specified very explicitly. Highly specified research objects limit the amount of available research data. This can be extremely problematic in the domain of formal sciences by causing irrelevant limitations. Data providers admit that some of the data request procedures in the past have been incoherent and lacked consistency. Still, the desire to support secondary EPR research exists within the data provider organization.

Data providers accent the purpose of defining use procedures primarily for the organization itself which affects the clarity of procedures concerning secondary use. For instance, technological solutions for receiving the actual EPR data to secondary use have not been specifically defined for all users. These solutions need to be defined individually depending on the use purpose, which harms the cost-effectiveness of the data request procedure. The process of gaining ethical permission for research is settled and does not include significant variance in this domain.

Data ownership and management issues affect significantly all secondary use cases in healthcare domain. In the context of this research, the hospital district which creates and stores EPR data is the legitimate data owner. Owner's responsibility is to define corresponding procedures for data delivery and choose potential candidates for secondary use. The technical implementation is outsourced to a company providing information technology services. If the data owner allows secondary use, the service provider is responsible for delivering the EPR data and charging corresponding costs.

On national level, healthcare organization's data warehouse is a crucial part of gaining access to EPR data in secondary purposes for both clinical researchers and product developers. Due to privacy concerns, business operators face limitations regarding the content of available EPR data. Properly implemented data warehouse provides a protected EPR processing environment, reduces the role of external service providers, and simplifies the data request procedure (see Fig. 1 "Protected processing" approach).

Privacy Preserving. Interview discussions indicate that the privacy preserving of unstructured free text in an EPR is challenging. Main discussion topics considered the level of privacy preserving mechanisms and user responsibilities in the context of secondary use of EPR data (see Fig. 1 "Privacy preserving").

Overall, the potential of free text content in EPR data is underestimated, and therefore not developed further. Procedures and mechanisms for structured EPR data are generally more advanced. Still, from the perspective of data provider, the basic privacy preserving principles for structured and unstructured content exist: secondary user's ethical responsibility and protecting patient's intimacy. Data providers can not monitor the secondary use, and therefore researchers and other secondary users need to be committed to following research ethics.

Data providers criticised the use of different privacy preserving mechanisms for non-trivial personal identifiers. It was commonly noted that unaltered data is more valuable for secondary use than privacy preserved data. Masking or removing all personal identifiers is extremely challenging for an automatic system, whereas manually performed privacy preserving involves a significant amount of labour. When considering free text in an EPR, protecting the data from unwanted use is preferred over using extensive privacy preserving mechanisms. Thus, the data provider favours the use of "Protected processing" approach introduced in Fig. 1. For research use, the extent of privacy related actions should be connected to the purpose of the research. The use of distorted data as a result of extensive privacy preserving is criticised. The increased level of privacy harms the value of unstructured data.

5   Conclusions

Possible secondary users and use perspectives concerning EPR data were discussed. Motivation to allow open access use of EPR data exists in the data provider organization. However, the lack of national open data related legislation and procedures limit the data availability in this domain. When considering EPR based product development, data providers and business operators have different views of the possibility to utilize data for this purpose. The process of gaining access to EPR data for secondary use included several factors restricting data availability. Data and information system ownership issues were found challenging. Service providers for EPR systems cause distraction in the data request process by managing the technical process between the data owner and possible secondary users. The use of privacy preserving techniques was found controversial. Since developing an automatic system to explicitly anonymize or de-identify EPR data is challenging or even impossible, the use of privacy preserving for non-trivial personal identifiers was criticised. In this context, restricted data processing environment was preferred over open access and extensive use of anonymization or de-identification techniques.

Based on these findings, two approaches to implement secondary use of EPR data were identified: data alteration and protected data processing. Both approaches provide privacy for corresponding patients: data alteration via the extensive use of privacy preserving and protected data processing via secure and exclusive processing environment. However, the impact on data availability and value is divergent for such approaches. The exclusive data processing environment is open only for limited amount of users with restricted use intentions,

Table 2. Approach specific effects

	Availability	Privacy	Data value
Altered data	+	+	−
Protected processing	−	+	+

further providing more valuable unaltered data. Data alteration offers relevant EPR data without strict boundaries, while decreasing the data value. A collection of approach specific effects is presented in Table 2.

References

1. Häyrinen, K., Saranto, K., Nykänen, P.: Definition, structure, content, use and impacts of electronic health records: a review of the research literature. Int. J. Med. Inform. **77**(5), 291–304 (2008)
2. Allvin, H., Carlsson, E., Dalianis, H., Danielsson-Ojala, R., Daudaravičius, V., Hassel, M., Kokkinakis, D., Lundgrén-Laine, H., Nilsson, G.H., Nytrø, Ø., et al.: Characteristics of Finnish and Swedish intensive care nursing narratives: a comparative analysis to support the development of clinical language technologies. J. Biomed. Semant. **2**(3), 1 (2011)
3. Leino-Kilpi, H., Välimäki, M., Arndt, M., Dassen, T., Gasull, M., Lemonidou, C., Scott, P., Bansemir, G., Cabrera, E., Papaevangelou, H., Mc Parland, J.: Patient's Autonomy, Privacy and Informed Consent. Biomedical and Health Research. IOS Press, Amsterdam (2000)
4. Neamatullah, I., Douglass, M.M., Li-wei, H.L., Reisner, A., Villarroel, M., Long, W.J., Szolovits, P., Moody, G.B., Mark, R.G., Clifford, G.D.: Automated de-identification of free-text medical records. BMC Med. Inform. Decis. Mak. **8**(1), 32 (2008)
5. Meystre, S.M., Friedlin, F.J., South, B.R., Shen, S., Samore, M.H.: Automatic de-identification of textual documents in the electronic health record: a review of recent research. BMC Med. Res. Methodol. **10**(1), 70 (2010)
6. Merkow, M., Breithaupt, J.: Information Security: Principles and Practices. Prentice Hall Security Series. Pearson Education, Upper Saddle River (2006)
7. Liu, C.H., Chung, Y.F., Chen, T.S., Wang, S.D.: The enhancement of security in healthcare information systems. J. Med. Syst. **36**(3), 1673–1688 (2012)
8. Furnell, S.: Securing Information and Communications Systems: Principles, Technologies, and Applications. Artech House Computer Security Series. Artech House, Boston (2008)
9. The Ministry of Social affairs and Health: Project STM011:00/2015 (2015). http:// stm.fi/hanke?selectedProjectId=6503 (Finnish)
10. The Ministry of Education and Culture of Finland: Open science and research (2016). http://openscience.fi
11. The European Parliament and the Council of the European Union: Directive 95/46/ec on the protection of individuals with regard to the processing of personal data and on the free movement of such data (1995). http://eur-lex.europa. eu/legal-content/EN/TXT/?uri=celex:31995L0046

12. Personal data act: (523/1999) (1999). http://www.finlex.fi/fi/laki/kaannokset/1999/en19990523.pdf
13. U.S. Department of Health and Human Services: Health Insurance Portability and Accountability Act (HIPAA) (2010). http://www.hhs.gov/hipaa/for-professionals/privacy/special-topics/de-identification/index.html
14. Safran, C., Bloomrosen, M., Hammond, W.E., Labkoff, S., Markel-Fox, S., Tang, P.C., Detmer, D.E.: Toward a national framework for the secondary use of health data: an American medical informatics association white paper. J. Am. Med. Inform. Assoc. 14(1), 1–9 (2007)
15. Ikitik: Information and language technology solutions for health information and communication (2016). http://www.ikitik.fi
16. Elo, S., Kyngäs, H.: The qualitative content analysis process. J. Adv. Nurs. 62(1), 107–115 (2008)

Is Home Telemonitoring Feasible in the Care of Chronic Diseases - Insights into Adherence to a Self-management Intervention in Renewing Health Finland Trial

Anna-Leena Vuorinen[1(✉)], Miikka Ermes[1], Tuula Karhula[2],
Katja Rääpysjärvi[2], and Jaakko Lähteenmäki[3]

[1] VTT Technical Research Centre of Finland, Tampere, Finland
anna-leena.vuorinen@vtt.fi
[2] Eksote South Karelia Social and Health Care District, Lappeenranta, Finland
[3] VTT Technical Research Centre of Finland, Espoo, Finland

Abstract. eHealth studies typically suffer from high attrition rates. **Objective** To investigate type 2 diabetes and heart disease patients' adherence to a self-management intervention that combined health coaching and telemonitoring. **Methods** Renewing Health Finland was a 12-month randomized controlled trial to improve quality of life (QoL) and/or HbA1c of 595 patients with chronic conditions. The intervention consisted of (1) weekly measurement of health parameters (2) health coaching every 4–6 weeks. Adherence to telemonitoring was defined as the percentage of weeks with at least one reported health measurement. Adherence to coaching was defined as the number of received calls. **Results** The median percentage of monitored weeks was 65 % without time-dependent attrition. 66 % of participants received 7–11 calls that corresponds to the predefined coaching schedule. Adherence did not correlate with QoL or HbA1c. **Discussion** Our results indicate that the intervention in the Renewing Health Finland trial was delivered almost with planned intensity.

Keywords: Adherence · Telemonitoring · Health coaching · QoL · HbA1c · Type 2 diabetes · Heart disease

1 Introduction

Increasing burden of chronic diseases combined with aging population and reductions in health care resources create a need to identify care models that respond to the growing demand while maintaining and/or improving the quality of the care.

Chronic care models that incorporate remote patient monitoring have shown promise in the past decades. Specifically telemonitoring interventions that involve measuring and reporting disease-specific information remotely and transferring the data to health care providers have been widely studied with positive outcomes among a variety of chronic conditions including diabetes, hypertension, chronic obstructive pulmonary disease and heart failure [1, 2]. Such interventions may improve patient's engagement with self-management while sharing the self-monitoring data with health care professionals provides continuous up-to-date information for care professionals to support

© Springer International Publishing Switzerland 2016
H. Li et al. (Eds.): WIS 2016, CCIS 636, pp. 57–66, 2016.
DOI: 10.1007/978-3-319-44672-1_6

their clinical decision making. However, there is a growing bulk of literature including large-scale rigorous randomized controlled trials that have failed to validate the hypothesized benefits of telemonitoring interventions [3–7].

Attrition is a common feature for eHealth studies; interventions that are not critical but are based on patients' voluntariness and that are easy to discontinue typically suffer from high nonusage rates [8, 9]. While it has been argued that high drop-out rates might be natural feature for eHealth trials, attrition is underreported and poorly understood and might even cause publication bias as it is often associated with failure of the intervention [8]. Study reports show aggregated statistics on usage, however, they often lack longitudinal aspects. Detailed analyses of uptake and possible discontinuation of the intervention are needed and required [10] to understand and further improve the effectiveness of eHealth interventions.

Between 2010–2013, we conducted a randomized controlled trial that assessed the effect of telemonitoring assisted self-management intervention on the quality of life and glycemic control of patients with chronic conditions [11]. Renewing Health Finland was a part of the European research project Renewing Health where nine collaborating countries assessed the effect of telehealth interventions with RCT settings. In conjunction with the majority of the other trials [5], Renewing Health Finland did not find significant effect on primary health outcomes. In this paper we seek to find potential factors that may have contributed to the nonsignificant findings by analysing patients' engagement with the intervention in the Renewing Health Finland trial and investigating the effect of adherence on the health outcomes.

2   Methods

2.1   Renewing Health Finland

Renewing Health Finland (RHF) was a 12-month randomized controlled trial aiming to improve health-related quality of life of type 2 diabetes and heart disease patients and glycemic control of type 2 diabetes patients with an intervention that combined telemonitoring and health coaching. The study population consisted of 308 heart disease patients (ischemic heart disease or heart failure) and 287 type 2 diabetes patients that were recruited from the health care district of South Karelia and further randomized either to the telemonitoring group or to standard care (2:1). Of the 595 participants, 361 (61 %) were men, mean age was 67 ± 9 years and mean BMI was 29.7 ± 5.2 kg/m^2.

The telemonitoring intervention consisted of weekly measurement of weight, blood pressure, blood glucose (for diabetics) and steps (heart disease patients) and reporting the measurements to the back-end systems using a mobile phone. Patients were given a self-monitoring toolbox that consisted of a mobile phone and measurement devices (incl. blood pressure meter, blood glucose meters, scale and a pedometers). In addition, each patient was assigned a personal health coach. During the first study visit the patient and the health coach created a self-management plan that included small achievable health behaviour changes agreed by the patient. The coach called the patients at a 4–6 week interval. During the calls the coach reviewed the goals and provided the patients with information, assistance and support to achieve them. Before each call, health coaches reviewed patient's self-monitoring data. If the data showed abnormalities, the coach advised patient to contact primary care. If self-measurement data were missing and the patient had received the self-monitoring devices, the coach reminded the patient of the importance of self-monitoring and asked to conduct self-measurements and report them to the system on a regular basis. Detailed description of the intervention and study results is described by Karhula et al. [11]. The telemonitoring system of Renewing Health Finland is depicted in Fig. 1.

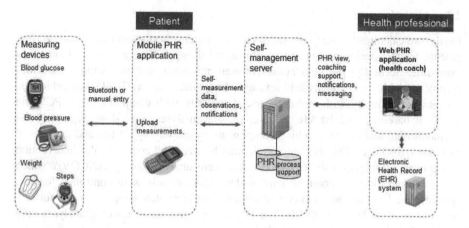

Fig. 1. Technical architecture of telemonitoring system used in Renewing Health Finland

2.2 Measures

The telemonitoring system stored time-stamped self-monitoring data originating from the patients' measurement devices. The log-data from coaching calls included starting time and ending time and status of the call (failed or answered). The logfiles allow detailed analysis of patients' engagement with the telemonitoring intervention in the course of time and the responsiveness to the health coaching.

Adherence. *Adherence to telemonitoring* was defined as the percentage of weeks that included at least one measurement (blood glucose, body weight, blood pressure, steps) reported to the system. Each week was analysed separately and was relative to the individual starting time. In addition, adherence to glucose monitoring was analysed separately as HbA1c was the primary outcome. Adherence to glucose monitoring was similarly defined as the percentage of weeks including at least one glucose measurement.

Adherence to health coaching was defined as the total number of answered health coaching calls and the mean duration of the answered calls. As coaching calls were planned to be done with a 4–6 week interval, each patient was supposed to receive 7–11 calls when the baseline and concluding calls were excluded. The first and last calls were excluded because the control group received the corresponding calls. However, the content of the calls was not similar but control patients received only study-related information.

Quality of Life. The primary outcomes of the Renewing Health Finland trial were the health-related quality of life (QoL) measured using SF-36 and HbA1c. SF-36 was collected at baseline and at the end of the study after 12 months. Physical component score (PCS) and mental component score (MCS) of SF-36 were used in the analyses.

HbA1c. Among type 2 diabetes patients the other primary outcome was HbA1c. Laboratory tests were done at baseline and at the end of the trial.

2.3   Analysis

Firstly, we calculated descriptive statistics for adherence to the overall telemonitoring, glucose telemonitoring and health coaching and illustrated the adherence trajectories over time. Secondly, the association between adherence and following sociodemographic variables were analysed: sex, age, body mass index (BMI), comorbidities, education, familiarity with mobile phone, familiarity with computer, QoL (PCS and MCS) at baseline, and HbA1c at baseline. The analyses were done separately for telemonitoring and health coaching components by employing a t-test and analysis of variance (ANOVA). Thirdly, the correlations between adherence to the intervention and primary outcomes (SF-36 and Hb1Ac) were analysed using ANCOVA which allowed controlling for potential confounders. QoL models were controlled for sex, age, baseline level of QoL and comorbidities. HbA1c models were controlled for sex, age, and baseline level of HbA1c. For HbA1c and glucose monitoring analyses only the type 2 diabetes group was included.

3   Results

3.1   Adherence to Telemonitoring

The majority of participants adhered to weekly telemonitoring (Fig. 2). The median percentage of adherent weeks was 65 %; the 25th and 75th percentiles were 27 % and 85 %, respectively. When only intervention completers were included, the median adherence increased slightly to 67 %. Of the 370 patients, 43 % and 29 % were 70 % and 80 % adherent with weekly telemonitoring. No major attrition was observed in the course of the 12-month follow-up. The retention with telemonitoring was fairly good with the percentage of adherent participants varying from 51 % (week 37) to 68 % (week 6). However, there was a statistically significant linear decrease over time ($\beta = -0.002$ [-0.003 to -0.001]). Adherence did not differ between the disease groups and the median number of monitored weeks was 30 in both groups.

Adherence to glucose monitoring among diabetes patients was investigated separately as HbA1c was the second primary outcome for the diabetes group (Fig. 3). Of the 180 diabetes patients, 39 (22 %) did not report any glucose measurements via the mobile application. The median number of glucose monitoring weeks was 17 (33 %). Twenty-six and eighteen percent of the patients were 70 % and 80 % compliant, respectively. Among those who started glucose monitoring, the median number of monitored weeks was 52. Similarly to the overall adherence to telemonitoring,

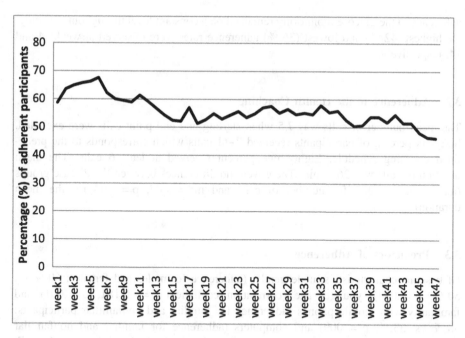

Fig. 2. Adherence to weekly telemonitoring - the percentage of participants who made at least one measurement on a given week (The figure is limited to the week 47 because trial's end-point visits were scheduled to start from week 48)

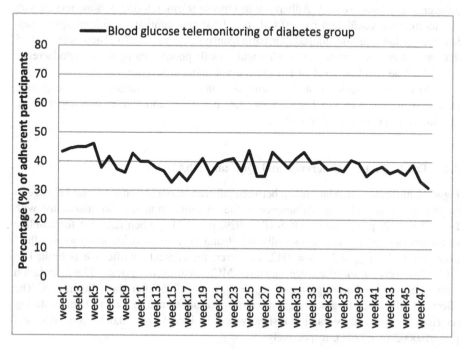

Fig. 3. Adherence to weekly telemonitoring of blood glucose; the percentage of participants who made at least one glucose report on a given week (The figure is limited to the week 47 because trial's end-point visits were scheduled to start from week 48)

adherence to the glucose monitoring remained on a constant level throughout the study; the highest (42 %) and lowest (36 %) adherence rates were observed at weeks 4 and 14, respectively.

3.2 Adherence to the Health Coaching

The mean number of calls was 7.5 when starting and end-point calls were excluded. Sixty-six percent of participants received 7–11 calls which corresponds to the predefined coaching schedule. Eighty-nine percent received at least 6 calls. The mean duration of calls was 26.7 min. There were no differences between the disease groups (md = .19, p = .305 for number of calls and md = .007, p = .993 for the mean duration).

3.3 Predictors of Adherence

Of baseline characteristics, male gender (adherence for males and females: 60 % vs. 50 %, p = .004), younger age (correlation coefficient r = −0.181, p < .001) and familiarity with mobile phones (adherence for familiar and nonfamiliar participants: 58 % vs. 33 %, p = .009) and computers (adherence for familiar and nonfamiliar participants: 63 % vs. 49 %, p<.001) were associated with higher adherence. In total, only 12 patients reported they were not familiar with mobile phones. Education level, BMI, baseline level of QoL and number of comorbidities were not associated with adherence to telemonitoring. Adherence to glucose telemonitoring was associated with age (correlation coefficient r = −0.144, p = .054) and familiarity with computer (adherence for familiar and nonfamiliar participants: 47 % vs. 28 %). There were only 5 diabetes patients who were not familiar with mobile phone; among those the adherence was 7 %. The baseline level of HbA1c did not affect adherence.

Adherence to health coaching was associated with familiarity with computer measured as a higher number (7.8 vs. 7.4, p = .027) and longer duration (27 vs. 26 min, p = .04) of coaching calls.

3.4 The Impact of Adherence and QoL and HbA1c

Figure 4 illustrates the relationship between adherence to telemonitoring and change in quality of life and HbA1c. Adherence to the telemonitoring was not associated with PCS (β = .012, p = .266) or MCS (β = .029, p = .071). When adjusted for baseline characteristics there was a statistically significant positive association between adherence and MCS: $\beta_{adj=}.033$, p = .042. However, the clinical significance is limited as 10 % unit increase in adherence increases MCS by only 0.3 points. The association with PCS remained nonsignificant β_{adj} = .011, p = .295 after the adjustments. The effects of adherence to glucose telemonitoring or adherence to overall telemonitoring on Hba1c were negative and statistically nonsignificant β = −0.068 (p = .695) and β = −0.0002 (p = .891), respectively.

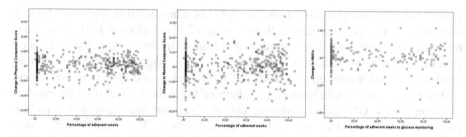

Fig. 4. The relationship between adherence to telemonitoring and change in quality of life

Health coaching measured as the number of answered coaching calls (β = .223, p = .551 for PCS, β = −.132, p = .808 for MCS) and mean duration of calls (β = −.049, p = .368 for PCS and β = .046, p = .557 for MCS) did not correlate with PCS or MCS.

4 Discussion

In this paper we investigated type 2 diabetes and heart disease patients' adherence to a self-management intervention that combined health coaching and telemonitoring, and further assessed the impact of adherence on QoL and Hba1c.

The health coaching component in the Renewing Health Finland (RHF) trial was realized closely as planned. Sixty-six percent (66 %) of patients received 7–11 coaching calls which corresponds to the predetermined 4–6week interval for health coaching, and 89 % of participants received six or more calls indicating the majority of participants missed one call at maximum.

Participants' adherence to the telemonitoring component was moderate: in the course of 12 months the participants conducted telemonitoring on 65 % of the weeks on average. Unlike in many eHealth interventions [8, 12] there was no major attrition towards the end of the trial but the percentage of adherent patients remained 51–68 % over time. In telemonitoring studies adherence rates have been shown to vary from 52 % to 75 % [2, 7, 13, 14]. However, in these studies likewise in a number of other telemonitoring trials, health parameters were monitored on a daily basis that makes the numbers incomparable. Moreover, adherence numbers are typically based on aggregated averages failing to take into account the time effect that was presented in our study.

Glucose telemonitoring was realized with lower adherence rates. Type 2 diabetes patients conducted glucose measurements on 33 % of the weeks on average and only 18 % of the participants were 80 % adherent. The results indicate lower engagement than found in other studies. For example in the DiaTel trial [14], 75 % of non-insulin type 2 diabetes patients engaged in daily glucose monitoring. Low adherence might be related to the unchanged HbA1c levels found in RHF study.

Interestingly, telemonitoring activity did not correlate with QoL or HbA1c but the changes in those health parameters were similar regardless whether a patient was high or low adherent to the telemonitoring component. Our results are contradictory to the earlier studies showing improved quality of life in telemonitoring trials [1, 15], and improved HbA1c as a result of self-monitoring of blood glucose [16], though the benefits among type 2 diabetics not using insulin are inconclusive [17]. The analysis of predictors of adherence did not shed light on what could explain the lack of this relationship. Participants' baseline BMI and HbA1c levels and QoL were not associated with adherence to telemonitoring. Adherence was neither affected by the number of comorbidities. Of baseline characteristics, sex, age and familiarity with a mobile phone and a computer predicted higher adherence. The results imply that technology-assisted interventions might have potential to appeal younger men who are typically underrepresented and nonadherent to lifestyle interventions. However, the clinical effectiveness of telemonitoring in this context remains questionable.

While self-monitoring is the cornerstone of effective self-management in chronic diseases, it is only as effective as actions taken in response to the measurements [18]. In RHF health coaching was designed to respond to individual needs and empower patients and educate patients for better self-management. Health coaches reviewed patients' self-monitoring data before each call, however, the medication changes that critically affect patient's condition, were not specifically addressed. We do not have data about whether coaching calls resulted in further actions to intensify pharmacological treatment.

5 Conclusions

Our results on adherence indicate that the intervention in Renewing Health Finland was realized with at least moderate intensity. The vast majority of participants received almost a complete health coaching intervention and the adherence to telemonitoring of health parameters was moderate with 34 monitored weeks on average. However, despite their engagement in the intervention, patients did not show improvement in QoL or glycemic control. In fact, the level of adherence to telemonitoring did not correlate with QoL or HbA1c at all. The results suggest that the self-management intervention components (telemonitoring and health coaching) were not effective in improving the quality of life and glycemic control of patients with chronic conditions even when successfully delivered and adhered to. Further research is needed to identify effective approaches to improve the care of chronic conditions.

Acknowledgements. This research received funding from TuTunKo-project which is supported by Tekes, the Finnish Funding Agency for Technology and Innovation and European Regional Development Fund ERDF.

References

1. Pare, G., Jaana, M., Sicotte, C.: Systematic review of home telemonitoring for chronic diseases: the evidence base. J. Am. Med. Inform. Assoc. **14**(3), 269–277 (2007)
2. Clark, R.A., Inglis, S.C., McAlister, F.A., Cleland, J.G.F., Stewart, S.: Telemonitoring or structured telephone support programmes for patients with chronic heart failure: systematic review and meta-analysis. BMJ **334**(7600), 942 (2007)

3. Steventon, A., Bardsley, M., Billings, J., Dixon, J., Doll, H., Hirani, S., Cartwright, M., Rixon, L., Knapp, M., Henderson, C., Rogers, A., Fitzpatrick, R., Hendy, J., Newman, S.: Effect of telehealth on use of secondary care and mortality: findings from the whole system demonstrator cluster randomised trial. BMJ **344**, e3874 (2012)

4. Cartwright, M., Hirani, S.P., Rixon, L., Beynon, M., Doll, H., Bower, P., Bardsley, M., Steventon, A., Knapp, M., Henderson, C., Rogers, A., Sanders, C., Fitzpatrick, R., Barlow, J., Newman, S.P.: Effect of telehealth on quality of life and psychological outcomes over 12 months (whole systems demonstrator telehealth questionnaire study): nested study of patient reported outcomes in a pragmatic, cluster randomised controlled trial. BMJ **346**, f653 (2013)

5. Kidholm, K., Stafylas, P., Kotzeva, A., Pedersen, C., Dafoulas, G., Scharf, I., Jensen, L., Lindberg, I., Andersen, A., Lange, M., Aletras, V., Fasterholdt, I., Stübin, M., d'Angelantonia, M., Ribu, L., Grottland, A., Greuel, M., Giannaokopoulos Isaksson, L., Orsama, A-L., Karhula, T., Mancin, S., Scavini, C., Dyrvig, A-K., Wanscher, C.: Regions of Europe working together for health FinalReport (2014). http://www.renewinghealth.eu/documents/28946/1008625/D1.12+v1.5+Renewing+Health+Final+Project+Report+-+Public.pdf. Accessed 04 Apr 2016

6. Chaudhry, S.I., Mattera, J.A., Curtis, J.P., Spertus, J.A., Herrin, J., Lin, Z., Phillips, C.O., Hodshon, B.V., Cooper, L.S., Krumholz, H.M.: Telemonitoring in patients with heart failure. NEJM **363**, 2301–2309 (2010)

7. Ong, M.K., Romano, P.S., Edgington, S., Aronow, H.U., Auerbach, A.D., Black, J.T., De Marco, T., Escarce, J.J., Evangelista, L.S., Hanna, B., Ganiats, T.G., Greenberg, B.H., Greenfield, S., Kaplan, S.H., Kimchi, A., Liu, H., Lombardo, D., Mangione, C.M., Sadeghi, B., Sadeghi, B., Sarrafzadeh, M., Tong, K., Fonarow, G.C.: Effectiveness of remote patient monitoring after discharge of hospitalized patients with heart failure: the better effectiveness after transition-heart failure (BEAT-HF) randomized clinical trial. JAMA Intern. Med. **176**(3), 310–318 (2016)

8. Eysenbach, G.: The law of attrition. J. Med. Internet Res. **7**(1), e11 (2005)

9. Wu, R.C., Delgado, D., Costigan, J., Maciver, J., Ross, H.: Pilot study of an Internet patient-physician communication tool for heart failure disease management. J. Med. Internet Res. **7**(1), e8 (2005)

10. Eysenbach, G.: CONSORT-EHEALTH: improving and standardizing evaluation reports of Web-based and mobile health interventions. J. Med. Internet Res. **13**(4), e126 (2011)

11. Karhula, T., Vuorinen, A.-L., Rääpysjärvi, K., Pakanen, M., Itkonen, P., Tepponen, M., Junno, U.-M., Jokinen, T., van Gils, M., Lähteenmäki, J., Kohtamäki, K., Saranummi, N.: Telemonitoring and mobile phone-based health coaching among finnish diabetic and heart disease patients: randomized controlled trial. J. Med. Internet Res. **17**(6), e153 (2015)

12. Mattila, E., Orsama, A.-L., Ahtinen, A., Hopsu, L., Leino, T., Korhonen, I.: Personal health technologies in employee health promotion: usage activity, usefulness, and health-related outcomes in a 1-year randomized controlled trial. JMIR mHealth uHealth **1**(2), e16 (2013)

13. Port, K., Palm, K., Viigimaa, M.: Daily usage and efficiency of remote home monitoring in hypertensive patients over a one-year period. J. Telemed. Telecare **11**(Suppl 1), 34–36 (2005)

14. Stone, R.A., Rao, R.H., Sevick, M.A., Cheng, C., Hough, L.J., Macpherson, D.S., Franko, C.M., Anglin, R.A., Obrosky, D.S., Derubertis, F.R.: Active care management supported by home telemonitoring in veterans with type 2 diabetes: the DiaTel randomized controlled trial. Diab. Care **33**(3), 478–484 (2010)

15. Inglis, S.C., Clark, R.A., McAlister, F.A., Ball, J., Lewinter, C., Cullington, D., Stewart, S., Cleland, J.G.: Structured telephone support or telemonitoring programmes for patients with chronic heart failure. Cochrane Database Syst. Rev. **8**, CD007228 (2010)

16. McAndrew, L., Schneider, S.H., Burns, E., Leventhal, H.: Does patient blood glucose monitoring improve diabetes control? A systematic review of the literature. Diab. Educ. **33**(6), 991–1011 (2007). discussion 1012–3, Jan
17. Malanda, U.L., Welschen, L.M.C., Riphagen, I.I., Dekker, J.M., Nijpels, G., Bot, S.D.M.: Self-monitoring of blood glucose in patients with type 2 diabetes mellitus who are not using insulin. Cochrane Database Syst. Rev. **1**, CD005060 (2012)
18. Kolb, H., Kempf, K., Martin, S., Stumvoll, M., Landgraf, R.: On what evidence-base do we recommend self-monitoring of blood glucose? Diab. Res. Clin. Pract. **87**(2), 150–156 (2010)

Welfare Issues of Children, Youth, Young Elderly and Seniors

Developing a Framework for App Evaluation: Empowering Learning and Communication with iPads for Children and Young People with Communication Impairments

Claire Hamshire[✉] and Julie Lachkovic

Manchester Metropolitan University, Manchester, UK
{c.hamshire,j.lackovic}@mmu.ac.uk

Abstract. This paper presents an overview of a two-year collaborative partnership project between Manchester Metropolitan University and three specialist schools in Manchester, UK. The purpose of the research was to work with a varied group of children and young people experiencing communication impairments and learning disabilities to gain an insight into their needs and subsequently to develop iPad based resources and an App evaluation framework for staff and carers supporting them.

The underpinning rationale of the project was to integrate iPads into the learning environment to provide opportunities for children and young people to further develop their skills and communication through the use of new technologies; where this was likely to enhance current practices. This paper outlines the successes and challenges raised by the project, as well as future developments and potential wider implications for iPad usage for this particular group.

Keywords: iPad · App · Evaluation framework · Communication · Learning

1 Introduction and Background

New and emerging technologies play an important role in enabling heath care providers and educationalists to respond to a range of challenges in the future [1, 2] and technology and innovation will increasingly influence the way that patients and staff perceive, understand and manage health and ill-health [3]. Staff from across health and social care professional groups increasingly use information and information communication technologies (ICT) for everything they do with patients, carers and members of the public [3] and tablet technologies have become a ubiquitous form of communication technology that have begun to permeate the way individuals learn, work and live [4].

People are adopting these new technologies at a rapid rate and tablet devices with a range of apps will become ever more of a key feature in improving the quality and cost of educating not just the formal workforce, but also students, patients and their carers [3]. eHealth will provide the backbone for the future citizen-centred healthcare environment [1] and as such, tablet devices and associate apps have the potential to transform future practice [3].

© Springer International Publishing Switzerland 2016
H. Li et al. (Eds.): WIS 2016, CCIS 636, pp. 69–81, 2016.
DOI: 10.1007/978-3-319-44672-1_7

When iPads were first launched in 2010, it was initially unclear as to the value of the device for promoting learning and teaching [5]. However, research suggested that tablet technology had a wide variety of applications that have the potential to enrich the learning experience and communication [6]. Tablet devices with associated apps are therefore being increasingly used within health and social care [3], and the National Health Service in the UK endorses a number of apps [7].

Technological interventions have often been used to assist in the learning of children and young people with communication impairments; with some evidence of success in the use of iPads to teach literacy [8–10]. In addition, health and teaching professionals consider iPads to be an important mode of technology that facilitates learning, communication and independence [11]. However, the speed of the implementation and uptake of these technologies is a workforce risk, as staff attempt to keep up-to-date with new technologies. There are currently no standardised guidelines relating to the use of tablets and apps in the UK, leading to variance and variety across practice and unresolved assessment and treatment issues; with the potential for patient complaints about disparities across services. Therefore, there was a need to develop a common framework and guidelines for practice, applicable to Health and Education staff who utilise tablet devices and apps to support patients and carers. Speech and Language Therapists were a key group of focus within this project as their role connects both health and education sectors.

2   Methods

This two-year longitudinal study was a collaborative partnership project between Manchester Metropolitan University (MMU) and three specialist schools in Manchester, United Kingdom: a nursery for children aged 3–5 years old; a special needs school for children aged 3–18 years old and a college for young people with additional needs aged 16–25. Together these schools provided education to children and young people with an age range of 3–25 years and a wide-range of additional learning and communication needs and diagnoses including Autism, Downs Syndrome and acquired head injury.

One of the challenges of using iPads with children and young people with communication and learning difficulties to support their learning and communication is searching through the large number of apps that are currently available and evaluating their usefulness in relation to a specific learning or communication need. Further, searches through the app store require exact names of specific apps for results to be generated and then choices between purchase or limited lite and free versions to be made. Therefore, the purpose of this research was to gain an in-depth understanding of the different elements of learning for these children and young people and subsequently develop an App evaluation framework to support future iPad usage and learning experiences in a way that promoted both social and personal as well as academic development.

The project team worked with educators, speech and language therapists, children and young people with communication impairments and their families to gain an insight into their needs and subsequently develop a portfolio of resources to support their learning and development.

3   Data Collection

Central to the whole project was the perspective of the children and young people, their parents and the school staff and we developed a dynamic, two-way process in a review of practice and evaluation of the technology. As such, the project drew on a phased, practice-driven and solution-focused action research approach [12] to make lasting change and provide a basis for good practice; developing a range of strategies and new innovations that addressed students' needs in more efficient ways. The aims of the project were:

1. To identify the specific issues and challenges that impacted on the use of iPads with children and young people with communication and learning difficulties.
2. To design practical suitable solutions that enabled staff and carers to assess the suitability of iPad and app usage for use across a wide spectrum of age, ability and need.
3. To develop training materials and linked training packages for teachers, carers and therapists.
4. To develop a compendium for prescribing iPad apps and linked training packages to provide a framework for the support of children and young people experiencing communication impairments and learning disabilities.

We needed to explore the perspectives of staff, carers and children and young people. In order to capture this data a multi-phase evaluation was undertaken with each distinct phase of the project conducted through a cyclical process of thinking, acting, data gathering and reflection [13].

In the first phase, we gathered information from staff and tutors at the schools by inviting them to participate in focus groups. In the second phase, we wanted to gain further insight into the students' perspectives as iPad users, as we believed that it was important to give the children and young people a 'voice' and involvement in their care and life choices [14]. Therefore, students with complex and multiple needs were invited to participate in an iPad user's group to test a selection of apps and give feedback on their experiences. In the third phase data from these focus groups was used to construct a framework for the assessment of an individual's 'tablet readiness' and an app evaluation matrix that was pilot-tested with children and young people at each of the schools before full implementation. In the final phase, the research team collated the individual app reviews and developed a compendium of case studies for use across the school sample sites along with an assessment framework.

4   Results

4.1   Phase 1 - Staff Perspectives

Data was collected via three staff focus groups, one at each of the three sample sites. The goal in these focus groups was to explore staff perceptions and experiences of using iPads and identify training and support needs. As such, there was no definitive schedule or set of questions; instead, staff were asked to complete an initial activity by writing comments individually on post-it notes to identify:

Perceived challenges and barriers to use
Level of technological confidence and ability

This feedback was then used to inform the group discussion activities. The data from the focus groups indicated that there was a range of prior experience and knowledge within the staff teams at each site. Whilst all teaching and support staff were interested in developing the use of iPads and apps some were concerned about their own technical skills when using the devices and the overall quality of some of the apps available to meet specific communication and learning needs. A significant challenge that was identified was that staff were unsure how to select appropriate apps for supporting particular learning needs and communication difficulties; particularly those that had limited prior experience of using tablet technology to support learning.

Sample written comments from the staff included:

'I have my own iPad and use a lot of apps already, I'm keen to get started'.
'I have no idea how to use one but with training I think I would be fine.'
'I feel that I would need further training as I feel that it could do so much more.'
'Unaware which applications are available or how to use them.'
'Not enough knowledge and not understanding how the iPad can be best used.'

When the research team collated the focus group feedback it was clear that a significant number of staff lacked confidence and basic skills to use the iPads therefore we developed a series of resources, linked to training sessions on:

how to assess individuals' abilities for iPad usage
how to set-up iPads to make them usable for children and young people with particular needs
adapting the iPad settings to enable individuals to use them effectively
evaluating and choosing iPad apps.

Following this training needs analysis the project team developed a series of both one-to-one and group training sessions and resources for staff to facilitate their skill development.

4.2   Phase 2 - Student's Perspectives

Data from students were collected from two of the three sample sites; as the children at the third site were under five and deemed too young to give in-depth feedback.

These student groups consisted of a total of twenty-one students with a range of abilities and communication needs - twelve male participants with Autistic Spectrum Conditions, four male participants with profound and complex learning needs and four female participants with profound and complex learning needs. The students were each provided with iPads and given a range of apps to use and asked to give feedback on a selection of apps. Those that were able to speak gave verbal feedback whilst symbols for: 'Fantastic', 'Not sure' or 'Terrible' were available for the non-verbal students.

All of the students indicated that they enjoyed using the iPads for a range of different purposes and the responses indicated that they particularly liked apps that allowed them to create images; record achievements using photos; sequence activities and use cause and effect apps to plan tasks. All of the apps were rated as 'fantastic' by the students using the symbols, sample comments from the students that were able to vocalise their perceptions included:

'I like it, I think the drawing is good.'
'I think it was special being able to use the iPads.'

4.3   Phase 3 – Development of the iPad Assessment Framework and App Evaluation Matrix

The next phase of the project had two specific purposes:

the development of an assessment framework to evaluate the needs of individual children and young people; and review whether iPad usage was possible or appropriate.
the development of an app evaluation framework so that a standardised review of the apps could be undertaken.

The project team began by working with individual children and young people to undertake an initial assessment to gain a shared understanding of their requirements and desires. The aim of this initial assessment was to work in partnership with families, carers and professionals to identify the most appropriate device, accessories (case, stylus, etc.) and apps that would support both communication and learning considering:

a typical range of postures/environments
the physical movements that an individual can complete
their visual processing of images
level of support available from family members and carers
usual method of communicating and interacting with others
the aims and expectations of iPad usage

Although each assessment was specific to the individual, common factors that needed to be considered were:

Cognitive abilities
Sensory abilities
Physical abilities
Environment
Support

The project team subsequently developed standardised guidelines relating to the use of tablets to ensure consistent assessment including:

an assessment framework for tablet management,
a guide on accessibility for tablet devices,

Following the development of these assessment criteria and guidelines the project team collaborated with school staff and therapists to identify core skills and curriculum areas over which the use of iPads would be evaluated within each of the three schools:

nursery for children aged 3–5 years old – cause and effect, mark-making,
special needs school for children aged 3–18 years old – mark-making, basic
numeracy and literacy
college for young people with additional needs aged 16–25 – cause and effect, art,
and music

Working in partnership with the school staff and therapists we subsequently
identified 100 apps for use across these curriculum areas. This was achieved via a
systematic search of the Apple iTunes app store; google searching for content areas e.g.
vocabulary, cause and effect as well as parent and practitioner recommendations. Our
focus was on the learning and communication needs of the children and young people
and we evaluated the apps' potential usefulness via a two-phase programme of testing.

For apps that were initially identified as potentially useful via the app store
descriptions, we developed a preliminary evaluation matrix that included the following
questions:

what specific activity/ability were you trying to encourage with the app?
was the app age developmentally appropriate for the child/young person?
was the app intuitive/accessible and easy to use?
does use of the app support individual needs and goals?
can the app be personalised for individual skill levels?
how does the app build on existing skills and support learning and/or
communication?
does the app allow the individual child/young person to do something new?

These preliminary evaluation criteria were informed by a review of the existing
school and therapy developmental assessment frameworks used to identify individual
needs at the three partnership schools, as well as app evaluation tools sourced via an
internet search. To test the usability and appropriateness of these criteria the research
team undertook a pilot test with three members of staff and six young people. The
reviewers in this pilot test gave both verbal and written feedback and these were
subsequently refined into:

guidance on how to evaluate tablet-based apps appropriately and select those that
can be used to meet specific needs,
fourteen-point evaluation matrix with space for open-responses (see Appendix 1).

4.4   Phase 4 – Development of the Compendium of Case Studies

Following the development of both the initial assessment tool for individual children
and young people and the app evaluation matrix, the final phase of the study was an
evaluation of the individual apps based on cumulative feedback from staff, parents and
children/young people using the double-sided form evaluation framework (see
Appendix 1).

Each of the school sample sites worked in partnership with the research team to
agree a list of apps that would be tested each school term. These apps were then tested

using an 'age and stage' approach within each of the three schools; whereby each app was used with children and young people with a range of both chronological and developmental ages as well as a range of specific learning and communication needs. As such each app was tested numerous times, across users and in a variety of different learning environments including class-based settings; community settings and home-settings.

This process of app testing allowed the research team to collate the reviews from the school staff, therapists and carers and identify where reviews were widely divergent. In the case of divergent reviews the research team undertook a second stage review to ensure that the assessment and review criteria had been applied consistently. If the criteria had not been applied correctly the reviews were discarded; if they had, then staff were interviewed to further explore their experiences.

At the end of each school term testing period, these reviews were subsequently developed into case studies. Each case study detailed a vignette of the use of an app for a specific user including, 'where', 'how' and the learning outcome achieved. These were uploaded on to an online database that the school staff used to inform future practice. The database was designed to facilitate finding apps via a range of routes, specifically users could search by:

developmental stage
- cooperative
- exploratory
- sensory
- sequences
- single action

learning goals
- cause and effect
- speech sounds
- taking turns
- vocabulary

curriculum areas
- early years
- communication
- mathematics
- music
- art

communication strategy
- language skills
- communication skills
- aided and unaided communication

On completion of the study 104 case studies were developed and made available to each of the schools to support further development of apps for learning and communication.

5 Discussion

This small-scale project at three specialist school settings within the UK provides an insight into how the use of iPads can be systematically evaluated and embedded within curricula to support learning and communication. By working collaboratively with school partners and responding to their concerns and challenges we developed a framework for both the assessment of iPad and app usage as well as a set of staff training materials.

Specifically, the aims of the project were to evaluate the issues and challenges on the use of iPads with this client group and design practical solutions that enabled staff and carers to support the learning and communication of children and young people with a wide-range of additional needs. During the initial staff focus groups some staff were positive about the potential of the iPads to support both student learning and communication. However a significant number also expressed concerns and an analysis of the focus group data identified that further training and support were required for staff to support usage of the devices. Through the design of training materials and linked training packages this project developed practical evaluation solutions that enabled staff and carers to assess the suitability of iPad and app usage for use across a wide spectrum of age, ability and need.

In common with previous research [15, 16], staff and students within this project found the introduction of a new technology into the curriculum to be of benefit. However, this study indicates that there is an on-going challenge for staff in keeping up-to-date with new and emerging technologies and in identifying and promoting the use of tablet devices and apps to facilitate learning and communication. Initial assessment of children and young people and systematic evaluation of apps is essential for appropriate usage of any tablet based technology, and whilst the results of this study are positive we note that the project focused on just three school settings and relied on reflective retrospective reports of the practitioners in those settings. As such this information is not directly generalisable to other settings but does provide an insight into the experiences of iPad usage within this group of children and young people, which may be transferable to schools with similar populations.

6 Conclusions

This paper, based on a dynamic, two-way process in a review of practice and evaluation of the technology, gives an insight into how systematic frameworks for iPad usage can be embedded into curricula to support learning and development. The potential for using iPads to improve learning outcomes for students with disabilities, complex needs and Autism Spectrum Conditions is currently unknown, and in order for iPad technologies to be harnessed as an effective learning tool consideration must be paid to the understanding, skill and confidence of those who support and teach the users. An increase in accessible training and support resources is needed to ensure that use of iPad technologies has an infra-structure to support success and negotiate initial technical challenges and to set appropriate expectations for both learners and adults involved in their care.

Each child/young person is an individual, technology can support individual needs and through the use of iPads and a selection of apps in this study we have enabled children to complete activities more independently and provide opportunities to enhance social and emotional growth. By utilising a combination of interdisciplinary expertise this research led to developments in assistive technology at each of the schools and supported learning, social inclusion and communication for the children and young people involved. Thus, by providing technology expertise and developing a systematic system of evaluating apps in partnership with both school staff and therapists we have turned the initial research into real-life benefits by developing a range of resources and initiatives to create opportunities for children's skill development and promote social inclusion.

The research has subsequently developed a basis for good practice for iPad usage with these children and young people; developing a range of strategies and new innovations that addressed students' communication and learning needs in more efficient ways. The interdisciplinary expertise arising from the project has also been used to develop a portfolio of educational, therapeutic and rehabilitative resources about tablet technologies, consultancy services and a compendium of apps appropriate for children with complex and multiple learning needs.

Acknowledgments. This study was part-funded by both Manchester Metropolitan University and Health Education North West.

The authors would like to acknowledge the contribution of each of the three specialist partnership schools and the Manchester Metropolitan University staff and students who contributed to the research project.

Appendix 1: App Evaluation Matrix

Name of reviewer ....
     Date   ....

Project   school
      ....

Name of app
      ....

Cost of app
      ....

Creator   of   the   app
      ....

Summary of the app content

Activities the app was used for/how are you using the app?

Age/abilities of student(s)/child(ren) using the app

Does the app allow you to do something you couldn't do before?

<table>
<tr><td></td></tr>
</table>

Please tick as appropriate

Content and components of the app	Yes	No	N/A
Authenticity – Does the app promote appropriate skill development?			
Differentiation – Does the app allow users to alter settings to meet individual needs?			
Usability – Can students/children use the app independently?			
Motivation – Are students/children motivated to use the app?			
Learning styles – Does the app address more than one learning style (visual/auditory/kinesthetic)?			
Images – Does the style of image add to the student/child's learning?			
Sound – Does the music/sound add to the student/child's learning?			
Balance – Does the app have a good balance of features to engage a range of users?			
Instructions – Are the instructions include with the app helpful?			
Support - Does the app have supporting information/website with additional information?			
Monitoring – Does the app allow for progress monitoring			

Data - Can student data be shared with teacher/parent?			
Accounts – Can more than one user account be set up?			
Interaction/communication – Does the app encourage the user to interact /communicate with others?			

Summary of the app - Using the space below please describe why you would or would not recommend the app

References

1. Department of Health (DoH): Embedding informatics in clinical education (eICE) (2014). http://www.eiceresources.org/
2. Haßler, B., Major, L., Hennessy, S.: Tablet use in schools: a critical review of the evidence for learning outcomes. J. Comput. Assist. Learn. **32**(2), 139–156 (2016)
3. Health Education England (HEE): Framework 15, Health Education England Strategic Framework 2014–2029 (2014)
4. McNaughton, D., Light, J.: The iPad and mobile technology revolution: benefits and challenges for individuals who require augmentative and alternative communication. Augmentative Altern. Commun. **29**(2), 107–116 (2013)
5. Berger, E.: The iPad: gadget or medical godsend? Ann. Emerg. Med. **56**(1), 21A–22A (2010)
6. Mang, C.F., Wardley, L.J.: Effective adoption of tablets in post- secondary education: recommendations based on a trial of iPads in university classes. J. Inf. Technol. Educ. Innovations Pract. **11**, 301–317 (2012)
7. NHS England: NHS Choices health apps library (2013). http://apps.nhs.uk/
8. Spooner, F., Ahlgrim-Delzell, L., Kemp-Inman, A., Wood, L.A.: Using an iPad2® with systematic instruction to teach shared stories for elementary-aged students with autism. Res. Pract. Persons Severe Disabil. **39**(1), 30–46 (2014)

9. Oakley, G., Howitt, C., Garwood, R., Durack, A.R.: Becoming multimodal authors: pre-service teachers' interventions to support young children with autism. Australas. J. Early Child. **38**(3), 86–96 (2013)
10. Johnson, G.M.: Using tablet computers with elementary school students with special needs: the practices and perceptions of special education teachers and teacher assistants. Can. J. Learn. Technol. **39**(4), n4 (2013)
11. Shane, H.C., Laubscher, E.H., Schlosser, R.W., Flynn, S., Sorce, J.F., Abramson, J.: Applying technology to visually support language and communication in individuals with autism spectrum disorders. J. Autism Dev. Disord. **42**(6), 1228–1235 (2012)
12. Todhunter, C.: Undertaking action research: negotiating the road ahead. Soc. Res. Update (34) (2001)
13. Savin-Baden, M., Howell Major, C.: Qualitative Research: The Essential Guide to Theory and Practice. Routledge, London (2013)
14. Coad, J., Flay, J., Aspinall, M., Bilverstone, B., Coxhead, E., Hones, B.: Evaluating the impact of involving young people in developing children's services in an acute hospital trust. J. Clin. Nurs. **17**, 3115–3122 (2008)
15. Ashburner, J., Ziviani, J., Pennington, A.: The introduction of keyboarding to children with autism spectrum disorders with handwriting difficulties: a help or a hindrance? Australas. J. Spec. Educ. **36**(01), 32–61 (2012)
16. Neely, L., Rispoli, M., Camargo, S., Davis, H., Boles, M.: The effect of instructional use of an iPad® on challenging behavior and academic engagement for two students with autism. Res. Autism Spectr. Disord. **7**(4), 509–516 (2013)

Pre-studies on Using Digital Games for the Elderly's Physical Activities

Aung Pyae[1(✉)], Mika Luimula[2], and Jouni Smed[1]

[1] University of Turku, Turku, Finland
{aung.pyae,jouni.smed}@utu.fi
[2] Turku University of Applied Sciences, Turku, Finland
mika.luimula@turkuamk.fi

Abstract. In this study, we conducted the pre-studies of Gamified Solutions in Healthcare (GSH) project. The main objectives of these pre-studies are to understand game design guidelines, the usability of Kinect, commercial, and non-commercial games. In pre-study 1, we conducted a literature review of motivational factors for the elderly's physical rehabilitation. In pre-study 2, we reviewed commercial games for the elderly. We conducted a usability testing of Microsoft Kinect with eight elderly in pre-study 3. Then, in pre-study 4, we conducted a usability testing of Puuha's game and two commercial games with five elderly participants. The findings from the literature review of motivational factors provide us useful game design guidelines how to motivate the elderly. The findings from the review of commercial games highlight that commercial games lack important design guidelines for the elderly. We also learn from the usability testing that Kinect has the potential to be used as an input device for the elderly, but built-in gestures are difficult for the elderly. Finally, the findings from the usability testing of Puuha's game and two commercial games provide us insightful game design and usability guidelines for the elderly. These findings are useful for our future game development for the elderly.

Keywords: Usability · Exergames · Gamification · Human-computer interaction

1 Background

Playing games can help the elderly to improve their quality of life, especially for the elderly who have often engaged in the leisure time and activities [1]. In recent years, there is an emerging trend of "intergenerational games", which involves a gameplay between the elderly and the younger generation [1]. The purpose of intergenerational games is to improve social connectedness between older and younger people as it can bridge the social gap between two age groups. One of the promising positive effects of the use of intergenerational games is that young people can have positive attitudes towards older adults when they have a tight social tie with seniors [2]. Moreover, it can prevent segregation between older and younger groups [3], and they can exchange strong social and emotional support each other [4]. There are other types of social games for the elderly such as online social games for the elderly [5], Nintendo Wii as an entertaining and socializing tool [6], and social exergames for the elderly [7].

© Springer International Publishing Switzerland 2016
H. Li et al. (Eds.): WIS 2016, CCIS 636, pp. 82–96, 2016.
DOI: 10.1007/978-3-319-44672-1_8

Digital games can be used to improve the physical fitness of the elderly. In stroke rehabilitation, therapists use digital games to improve patient's physical strength to regain mobility and to overcome functional deficits of upper and lower limbs. It can be used as an alternative tool to motivate the elderly to be more engaged in the regular exercise routines which are generally monotonous. Pyae et al. [8] designed and developed an augmented reality based rehabilitation games that include Activities of Daily Living (ADLs) games for the elderly stroke patients to train them to regain functional ability. They tested the game system with the elderly patients and report that elderly patients were more motivated and engaged in the therapeutic training after playing two-week game-based sessions. Marinelli and Rogers [9] advocate that exergames can help the elderly to improve their levels of physical activities that lead to positive health benefits. Kahlbaugha et al. [10] investigated the effects of compensatory strategies that were offered by Nintendo Wii on the physical activity, loneliness, and mood of the elderly. They report that the elderly group who played Wii during the ten-week study had lower loneliness and a greater positive mood. There are other types of game applications that are targeted for the elderly's physical activities such as Nintendo Wii for the elderly [6], Dance Dance Revolution (DDR) system [11], and robot games for the elderly [12].

Regarding digital games as recreational activities for the elderly, Pyae et al. [13] point out that leisure activities such as playing board games, physical-based digital games (e.g., Wii Bowling), singing songs, participation in social events and activities, and shopping, are regarded as motivational factors for the elderly in their older life. By using latest technologies, people design digital games that can enhance the elderly's experiences in doing leisure activities.

In our Gamified Solutions in Healthcare (GSH) project, the core concept is "Virtual Nursing Home", which consists of four services: Socialization, Rehabilitation, Entertainment, and Counseling respectively. In this study, we focus only on "Rehabilitation" service, which aims at promoting the elderly's physical well-being by utilizing game-based exercises. Before design and development of effective game-based exercises for the elderly, we conducted four pre-studies: a literature review, a review of commercial games, a pilot usability testing of Kinect, and a pilot usability testing of existing games. In the next sections, we will report the findings from these four pre-studies.

2 Pre-study 1: A Literature Review

In this literature review, we investigated the important factors that can influence on the motivation level of the elderly. Social functioning is regarded as one of the most important factors for the elderly's motivation [14–17]. When we design games for the elderly, we can create multiplayer games, intergenerational games, and networked games to improve their socialization. The relationship between therapist and the elderly is vital. It can have a great influence on patient's level of motivation [14, 18]. In designing games, we can create a virtual therapist or virtual caregiver in the game to communicate with the elderly. Having a personal goal in older life is one of the motivational factors [19]. We can create a game that has a goal-oriented level design to attract players from novice to professional levels.

Rehabilitative environment is an important factor for the elderly's motivation [20]. Thus, when we design digital games for physical rehabilitation, we can design elderly-friendly game context (e.g., household environment, park, and shopping mall). In physical rehabilitation, the individual motoric level can be different from one player to another. That is why, customization is important in rehabilitation [21]. When we create digital games for the elderly's rehabilitation, we need to take into account that customization is necessary so that they can meet their individual needs in playing games. There are other important motivational factors for the elderly and their rehabilitative activities such as meaningful rehabilitative tasks [22], information from healthcare professional [22], positive feedback and encouragement from therapist [23], music for rehabilitation [24], and recreational and leisure activities [25]. All the motivational factors are useful and insightful for game design and development. The detailed discussion of this section is mentioned in our previous study [13]. Table 1 shows the motivational factors for the elderly and proposed game design in developing motivation-driven games for the elderly.

Table 1. Motivational factors and game design.

Motivational factors	Game design consideration
Social functioning	- Multiplayer game - Intergenerational game - Virtual friend - Video chat in game - Social-networking game
Patient-therapist Relationship	- Virtual therapist - Virtual nurse - Virtual Coach
Goal setting	- Level design - Perceivable and achievable goals
Rehabilitative setting and environment	- Game theme and scenery - Difficulty of the game - Complexity of the game
Information from Healthcare Professionals	- Game tutorials - Game introduction - Help system - Computer-controlled assistant

3   Pre-study 2: A Review of Commercial Games

In this pre-study 2, we review existing games available in the market and research. First of all, we found a number of digital games in the market that support multiplayer gameplay. For instance, Nintendo Wii Tennis supports multiplayer gameplay in which two players can play the game together. In this way, the elderly can not only play the game but also socialize through the gameplay. Moreover, the Wii Sports games are promising for the intergenerational game and it is suitable for the elderly to play. Theng et al. [6] study the effects of Nintendo Wii games as an entertaining and socializing tool for improving the mental and social well-being of the elderly. They conducted a pilot study with 14 pairs of elderly-teenager participants, and reported that Wii games can bridge the inter-generational gap between older and young people.

The idea of virtual coach for physical exercises can be seen in the Nintendo Wii's My Fitness Coach game. In this game, the virtual personal fitness coach will guide the player through dynamic workouts for physical exercises, and the player can customize the workout plan, goals, environment, and music. Another type of virtual trainer game can be seen in Microsoft Xbox Fitness games, where virtual trainers are the real persons so that the player can socially connect to the trainer in the game.

With respect to setting a relevant rehabilitative goal, this concept can be also seen in Nintendo Wii's My Fitness Coach game. In this game, player can set his or her workout goal, calendar, type of exercises, and level of difficulty. There are a few goal-oriented digital games in the market. For instance, in Nintendo Wii's Dance Dance Revolution provides goal-based challenges, and it allows players to earn game scores or points that can unlock new costumes and songs in the game itself.

The familiarity with the rehabilitative setting and environment is one of the most important factors for the elderly's motivation. In the gaming market, there are a number of physical game-based exercises that support elderly friendly and real-world environment to players such as Nintendo Wii's My Fitness Coach game, Wii Sports game, Dance Dance Revolution game, Microsoft Xbox Fitness Game, Nike + Kinect Training, and VirtualRehab games.

Regarding information from healthcare professionals, there are also a number of digital games that support relevant information or feedback to players. For example, in PlayStation's EyeToy: Kinetic Fitness games the virtual coach or instructor gives the relevant information or instruction to players. In addition, there are other games that provide relevant information or instruction from virtual instructor such as Nintendo's Wii Sports, Microsoft Xbox Fitness game, and VirtualRehab games.

Real-world activity games are likely to enhance players' motivation and engagement. There are many games in the market that base on the real-world activities such as physical sports (soccer, tennis, and boxing), household activities (cooking and cleaning), recreational activities (shopping and playing cards), and cognitive activities (chess and puzzles). For example, Nintendo Wii' Sports games, Microsoft Xbox Fitness games, and PlayStation's EyeToy support game-based sport activities games are quite successful in the market and among younger players. For household activities, there are a few numbers of games available in the market such as Cooking MaMa and VirtualRehab's Window Cleaning Game.

Recreational activities games (e.g.,Virtual Shopping Games for Rehabilitation, Microsoft Solitaire Collection, Microsoft Xbox Karaoke Revolution) and cognitive games (e.g., Candy Crush and Microsoft Chess Titans) can be seen in the market. Moreover, there are similar games available in research projects by different institutions and organizations.

The game customization or pre-setting is one of the most important game principles and many digital games apply this principle in their games. The leading games in the market such as Nintendo Wii, Microsoft Xbox, PlayStation EyeToy, and research-based VirtualRehab games provide customization in the game. Apart from VirtualRehab games, most of the games available in the market are targeted for healthy and young players. Therefore, the customization of these games is not elderly-friendly.

With regard to the positive feedback for the elderly, most of the digital games available in the market support normal feedback to players such as scores, rewards, incentives, player's lifespan, and virtual cash. There is no or limited games that have encouraging and positive feedback to players regardless of their performance.

Most of the games in the market use appealing background music and audio feedback to players to be more motivated and engaged in the game itself. However, not many games support personalized music for players. Some of the musical games are quite successful in the market such as Microsoft Xbox Karaoke Revolution, Nintendo Wii's Music, and PlayStation Guitar Hero.

With regard to recreational activities, as mentioned earlier in this section, there are a few numbers of digital games that support recreational activities for players such as Wii Sports, Xbox Fitness games, Dance Dance Revolution, Microsoft Card games, and Karaoke Revolution game.

Based on this review, we found out that most of the games available in the market are not designed for the elderly and their physical rehabilitation. Although there are a number of games that have the potential for the elderly's physical rehabilitation, they still lack proper game designs to motivate the elderly. For example, Nintendo Wii's Sports games have the potential to train the elderly to improve their movements in rehabilitation but these games are not well-designed in customization that can be adjusted according to the elderly's ability in movements, range-of-motions, and strength. The findings from this literature review study highlight the design issues of commercial games in the market. It also helps us to understand insightful game design guidelines that are important for our future development phase.

4   Pre-study 3: A Pilot Usability Testing of Kinect

4.1   Kinect for Healthcare Games

Multimodal input devices are as important as game design and gameplay for the elderly. There are various types of multimodal input devices in the market and research such as gesture-based input device (e.g., Microsoft's Kinect, ASUS's Xtion), controller-based input device (e.g., Wii Remote, PS3's PlayMove), traditional controller (e.g., Xbox 360 controller), touch-based interaction (e.g., Touch-based monitor, touch-based mobile, and tablet), and robot-based interactive equipment (e.g., assistive technology for rehabilitation). Among them, Microsoft's Kinect draws a large attention from developers, researchers, and healthcare professionals because of its controller-free interaction. Kinect seems to be an effective way of interacting with system especially for elderly because the user does not need to hold a particular device or controller and it can reduce the cognitive load of the user.

According to Microsoft's Xbox [25], the new Kinect includes the new features such as giving a command Xbox and TV with the voice and gestures, playing games where the player is the controller, calling friends and family with integrated Skype in HD, broadcasting gameplay live with pictures, and being recognized and signed-in automatically. Since the first generation of Kinect was introduced, researchers and developers are trying to utilize Kinect in many applications (e.g., games) to help the ageing society in terms of their socialization, rehabilitation, safety, communication, and recreation. By using Kinect's features, the elderly can play physical games to improve their mobility and movement and communicate with friends and family members through Skype feature. Furthermore, Kinect is used as a monitoring system to detect the elderly's movements at home or senior homes (e.g., fall prevention and alert system). Kinect is also used as a recreational platform for the elderly to do their recreational activities (e.g., dance games).

According to Pyae et al. [13], they report that Kinect is an efficient and effective way of interaction for the elderly to play the rehabilitative games. Pisan et al. [26] mention that gamified exercises using Kinect to track player's lower limbs can encourage the elderly to be more engaged in the exercise regimes [27]. Webster and Celik [27] points out that Kinect is the front-runner in the market because it provides the natural movements of human body and it is the most feasible technology for exergaming.

Generally, Kinect is financially affordable and medically beneficial to the ageing society [28]. According to Ganesan and Anthony [28], they developed a Kinect-based physical exercise program to improve elderly's level of motivation in doing exercises. They conducted a usability testing with elderly participants and reported that the results were promising. Pompeu et al. [29] report that Kinect-based training for people with Parkinson's disease was safe and feasible for the patients, and they have improved in balance and gait activities after going fourteen 60 min sessions. In this study, we investigated the usability of Kinect for Xbox for the elderly. The details of usability testing and the findings are reported in the next sections.

4.2 A Pilot Usability Testing of Kinect

We conducted a pilot usability testing with eight elderly participants who reside at the service homes in Finland. The main objective of this usability testing is to investigate the usability of Kinect for Xbox as well as to understand the elderly's user experiences. Firstly, we recruited elderly participants through the advertisement at the center. The age range of the elderly participants is aged between 64 and 78. Most of them are currently residing at the service home, whereas the rest are the regular visitors to do their physical and social activities. There were equal numbers of male and female participants. Generally, elderly participants are physically and mentally sound and stable. They do physical exercises regularly and often involve in the social activities at the center.

According to the pre-survey, half of them spend at least 3 h per week for doing physical exercises while the other half spend more than 6 h per week. In this study, the elderly participants need to use their hand movements to control the computer system based on the different actions (e.g., wave to Kinect and swipe). Based on the pre-survey, all of them are right-handed. They have prior experiences in using TV remote control, whereas 2 out of 8 participants are familiar with computer tablets and smartphone. The only four participants use personal computers. None of them have used digital game consoles or smart TV.

In the usability testing, it included one Kinect-based tutorial that includes ten gesture-based tasks. Every participant needed to perform every task in the tutorial by using their hand. For example, the participants wave to the Kinect sensor and find the target by moving their hands. Before they performed each gesture-based task, the researcher demonstrated them how to perform the task. After a participant has finished a particular task, the researcher evaluated their performance by giving the different level of success. It includes "Independently and easily (4)", "Independently after a little training (3)", "With a tester's additional help (2)", and "The mission did not succeed (1)". If the elderly participant cannot perform a particular task, we skipped to the next task. The complete tutorial and description are shown in Table 2.

Table 2. Kinect-based tasks.

Task	Description
Wave to Kinect	Move your forearm left and right
Find a target	Point the palm of your hand toward the screen, and move your palm so the hand-cursor on a screen moves over the image
Make a selection	Move your hand over a specific item or tile
Scroll through a screen	When the hand icon appears on the screen, close your hand anywhere over the area you want to scroll. Move your hand to the left or right to pull the screen in that direction
Return to home	Hold out both of your hands towards the edge of the screen. Close your hands and move them toward each other in front of your chest
Return back to tutorial	Hold your closed hands toward each other in front of your chest Move your closed hands towards the edge of the screen and in the end, open your hands
Zoom in	When the hand icon appears on the screen, close your hand over the area you want to zoom Pull your hand toward you to zoom the screen in
Zoom out	When the hand icon appears on the screen, close your hand over the area you want to zoom Push your hand away from your body to zoom the screen out
Open the system menu -skype	Move the hand cursor over the Skype button Extend your arm forward toward the Kinect and hold until a circle timer appears After the ring has filled, you'll see the system menu. Select Pin to Home from the menu like you selected the application
Open the system menu -explorer	Move the hand cursor over the Skype button Extend your arm forward toward the Kinect and hold until a circle timer appears After the ring has filled, you'll see the system menu. Select Snap from the menu

After each participant has done the Kinect tutorial, we conducted a post-tutorial survey that contains the usability questionnaires to get the feedback from the elderly participants. Basically, the usability questionnaires are divided into two parts: usability of Kinect sensor and the participant's self-efficacy in using this device. The questionnaires use 5-point Likert scale from "Strongly disagree (1)" to "Strongly agree (5)".

The feedback given by the elderly participants were recorded by the researcher. Then, we have conducted a follow-up interview session with all participants to understand their user experiences during the tutorial session. Table 3 shows the detailed procedures of the Kinect tutorial session.

Table 3. The design and procedure of Kinect tutorial.

Activity	Duration
Introduction and pre-study interview	10 min
Kinect tutorial tasks 1–10	20 min
Post-study usability and self-efficacy questionnaires	10 min
Post-study interview	5 min
Total	45 min
Introduction and pre-study interview	10 min

4.3   Findings and Discussion

Firstly, the maximum number of participants who have succeeded a particular task without support from the researcher is five. For example, there are only five participants who can independently and easily perform the task "Wave to Kinect". According to Table 4, we can see that there are only three tasks succeeded by five elderly participants ("Wave to Kinect", "Return to Home", and "Return back to tutorial"). In contrast, the task called "Open the system menu-Explorer" has only one participant who has succeeded, followed by the task called "Open the system menu –Skype" that was succeeded by two participants. The tasks that have the maximum number of participants who have failed the mission are "Make a selection", "Zoom out", "Open the system menu-Skype", and "Open the system menu -Explorer" respectively. Table 4 shows the participants' level of success in the tutorial. We used the numbers to reveal the different levels of success by the elderly participants and the description of each number is shown below Table 4.

Table 4. The participants' level of success.

Task	P1	P2	P3	P4	P5	P6	P7	P8	Mean
Wave to Kinect	4	4	2	3	4	4	4	3	3.5
Find a target	4	4	3	2	4	4	1	1	2.8
Make a selection	4	4	1	1	4	4	1	1	2.5
Scroll through a screen	2	2	2	2	4	4	1	1	2.2
Return to home	4	4	4	4	4	3	1	1	3.1
Return back to tutorial	4	4	4	4	4	3	1	1	3.1
Zoom in	4	4	2	1	4	3	1	1	2.5
Zoom out	4	4	1	1	4	3	1	1	2.3
Open the system menu -Skype	3	4	1	1	4	3	1	1	2.2
Open the system menu -explorer	3	4	1	1	3	3	1	1	2.1

1- Mission did not succeed 2- With a tester's additional help
3- Independently after a little training 4- Independently and easily

Table 5. Average score.

Kinect task	Mean	SD
Wave to Kinect	3.5	0.7
Find a target	2.8	1.3
Make a selection	2.5	1.6
Scroll through a screen	2.2	1.1
Return to home	3.1	1.3
Return back to tutorial	3.1	1.3
Zoom in	2.5	1.4
Zoom out	2.3	1.5
Open the system menu -Skype	2.2	1.3
Open the system menu -explorer	2.1	1.2

According to the data shown in Table 5, "Wave to Kinect" is the easiest among other tasks in this tutorial (M = 3.5, SD = 0.7). The second easiest task for the participants are "Return to home" and "Return back to tutorial" (M = 3.1, SD = 1.3) respectively. In this tutorial, only "Wave to Kinect" is the one which was succeeded by all elderly participants, whereas the other tasks have the failed attempts by at least two up to four elderly participants. The scrolling task (M = 2.2, SD = 1.1) and opening the system menu tasks (M = 2.2, SD = 1.3 for Skype and M = 2.1, SD = 1.2 for Explorer) are the hardest tasks for the elderly participants. The tasks "Make a selection" (M = 2.5), Zoom in (M = 2.5), and Zoom out (M = 2.3, SD = 1.6) have the average success rate. With regard to the individual participants, it can be seen that Participant 5 (P5) has the highest level of success in most of the tasks. The participant 2 (P2) also has the second highest level of success, which is slightly lesser than the participant 5 (P5). The participants (P1, P2, P5, and P6) have no failed attempts in this tutorial. The participant 3 (P3) has the least number of failed attempts, whereas the participants (P6 and P7) have the highest number of failed attempts. In other words, P6 and P7 could only succeed a single task "Wave to Kinect". Table 5 shows the average score and standard deviation of each task in this tutorial.

In this tutorial, only half of the participants could succeed all the given tasks. By looking at the pre-interview data, it can be seen that the participants (P1, P2, P5, and P6) who succeeded all tasks have the prior experiences in using personal computers, tablets, and smartphones. However, there is no evidence seen in the findings whether their prior experiences can influence on this performance in this tutorial.

Regarding the usability of Kinect sensor, we consolidated the data and calculated the mean score and standard deviation (See Table 6). The overall results show that the usability of Kinect is just an average for the elderly participants. The questionnaire (Q7) has the highest mean score (M = 3.62, SD = 0.91), whereas the questionnaire (Q3) has the lowest (M = 2.75, SD = 1.16) followed by the questionnaire 2 (M = 2.87, SD = 0.99). The rest of the questionnaires have achieved equal or greater than 3. In general, most of the elderly participants think that the usability of Kinect is a challenge for them. However, they mention that they are confident to use it and they will be able to learn it through adequate training. They reveal their interests in using this sensor in interacting with computers. Table 6 shows the average scores of the usability of Kinect.

Table 6. Usability of Kinect.

No	Questionnaire	Mean	SD
1	I believe that I would use this kind of interactive device regularly	3.25	1.16
2	In my opinion, this interactive device was overly complicated	2.87	0.99
3	In my opinion, using the device was easy	2.75	1.16
4	I suppose that I might need help with using the device from a person who knows well technical devices	3.0	1.51
5	In my opinion, many features of the device were well integrated	3.0	0
6	In my opinion, there were too many inconsistencies in this interactive device	3.0	1.06
7	I believe that most of the users would learn to use this kind of device very quickly	3.62	0.91
8	In my opinion, this device is difficult to use.	3.25	1.16
9	I am confident to use this device	3.50	0.75
10	I should learn a lot of things until I could use this kind of interactive device	3.12	1.12

With regard to their self-efficacy in using this device, it can be seen that most of the elderly participants were confident to use it. The questionnaires regarding self-efficacy have achieved the mean score greater than 3 and the highest score is (M = 3.87, SD = 0.99) in the questionnaires (Q4 and Q5), whereas the lowest score can be seen in the questionnaire 3 (M = 3.0, SD = 1.51). Table 7 shows the average scores of elderly's self-efficacy towards using Kinect.

Table 7. Elderly's self-efficacy.

No	Questionnaire	Mean	SD
1	I am sure that I accomplished the assignments effectively	3.12	1.35
2	Although the assignments were challenging, I did pretty well	3.5	0.75
3	I succeeded in winning many challenges during performing the assignments	3	1.51
4	I believed that I could perform the given assignments also when I face difficulties	3.87	0.99
5	Generally speaking, I believe that I can achieve great and important results	3.87	0.99

According to the post-tutorial interview session, half of the elderly participants mentioned that they encountered difficulties in using Kinect sensor, whereas the other half had no significant issues. The elderly participants who succeeded all the given tasks in this tutorial mentioned that they enjoyed using the device and would like to use regularly in playing games and interacting with the computer. The two participants (P6 and P7) mentioned that they were frustrated to use it. The participants (P4 and P5) revealed that they are interested in learning how to use it although they had some failures in the tutorial.

This Kinect usability testing session highlights some important findings. Firstly, the elderly participants who have prior experiences in using computer, tablets, and smartphones can efficiently use the Kinect sensor. However, there is no evidence of a relationship between their prior experiences and their user experiences in using Kinect. Secondly, the majority of the elderly participants had succeeded the tutorial to a certain extent. They have also mentioned their interests towards the use of Kinect in future play. Thirdly, the majority of the elderly participants have mentioned their confidence towards using Kinect sensor in interacting with a computer. More importantly, they mention that they would be able to learn quickly how to use it through adequate learning. Moreover, we have learned that some built-in hand gestures of Kinect for Xbox in this tutorial are difficult for the elderly participants, whereas some gestures are easy for them. The other finding in our usability testing is that the software system that we have used in this tutorial is not well-designed for the elderly. Hence, there might have some usability issues for the elderly to control the Kinect sensor. From this study, we can learn that it is important to design user-friendly gestures when we use Kinect sensor for the elderly. The limitation of our study is that the sample size is small.

5 Pre-study 4: A Pilot Usability Testing of Existing Games

5.1 A Pilot Usability Testing

In this study, we conducted a pilot usability testing of existing games, which include Puuha's SportWall game and two commercial games: Xbox's Climbing game and PlayStation's Tennis game. The main objective of this study is to understand the elderly's feedback and user experiences in playing SportWall game and commercial games. SportWall game is designed and developed by Puuha Group Finland, and it is targeted for the elderly's physical exercise by using game-based physical activity. This game is simply designed based on therapeutic actions such as side-swaying, sit-to-stand, and light jump. This game is implemented using Unity3D game engine, and it uses a simple webcam, which uses Xtreme Reality technology to detect player's movements. Regarding commercial games, we select Microsoft Xbox's Climbing game and PlayStation's Tennis game because Xbox's Kinect-based Fitness games and PlayStation's controller-based PlayMove Fitness games provide a variety of exercise games. However, these games are not designed for the elderly. Thus, in this study, we would like to investigate if commercially available games are suitable for the elderly and their physical activities. Figure 1 shows a screenshot of Puuha's SportWall game.

The usability testing took place at one of the elderly service homes in Finland. We recruited five elderly participants for this pilot study. Basically, elderly participants in this study are 60 years old and above, and they are physically and mentally stable. They are the regular visitors to the elderly service home. Furthermore, they need to tolerate for 10 min in standing position to play the games. Before the elderly played the games, the researcher asked pre-interview questions. After that, the elderly needs to go through a game tutorial session to play a particular game. There were three game stations for three games: SportWall, Climbing, and Tennis games respectively. Elderly participants were randomly assigned to each game station. After the elderly has gone through a

Fig. 1. Puuha's game

game tutorial session, he or she played a particular game for five minutes, followed by a questionnaire session in which the researcher asked the elderly's user experiences in the gameplay and the usability of the game. The elderly had to go through the same procedure for next two games. After finishing all three games, the researcher asked post-game general interview questions that include the elderly's overall experiences in playing three different games. The whole session would require an hour for the individual elderly participant. Figure 2 shows a photo from the usability testing of existing games.

Fig. 2. Usability testing

5.2   Findings and Discussion

The findings show that commercial games are not user-friendly for the elderly. We investigated that the user interfaces, game contexts and contents, and gameplay in the commercial games are not suitable for the elderly. In Xbox's Climbing game, the elderly were distracted by the game contents such as user interface, icons, feedback, and game audios. Due to the complex game interface, it affected the elderly's gameplay and sometimes, they did not know how to continue their gameplay. In PlayStation's

Tennis game, the elderly found out that it was the hardest to play because of the PlayMove game controller. Furthermore, the elderly faced some challenges in playing the game because of the cluttered interface, contexts, and contents. Among three physical activity games, SportWall game was well-accepted by the elderly participants because of its simple, uncluttered, and clear game interface, contexts, and contents. Moreover, the gameplay in SportWall game is simple enough for the elderly to pay attention to their gameplay. According to the elderly's feedback towards the games, we found out that SportWall game is the most effective game for the elderly, followed by Xbox's Climbing game in the second place and PlayStation's Tennis game in the third place respectively.

Regarding the interactive input devices, we used three different multimodal input devices: Microsoft Kinect sensor, PlayStation's PlayMove controller, and traditional webcam. Among them, based on the feedback from the elderly participants, Microsoft Kinect sensor is the most effective device to play the game, followed by traditional webcam. The elderly participants gave negative feedback towards PlayStation's PlayMove controller that it is too complicated for them to control the game. Basically, the elderly needs to press particular buttons on the controller and sometimes, they forgot to press the buttons and as a result, they could not continue the gameplay. Based on the findings from the post-game interview sessions, the elderly participants mentioned that they had fun in playing games except the fact that they encountered some difficulties in some games. Moreover, they are interested in playing these games again and advocated that they can improve their ability in playing games through adequate training and guidance from a trainer. In general, they advocated that digital games seem to be an effective way of exercising for their physical well-being.

Based on the findings from this pilot study, we summarize the following usability and game design guidelines for our future game development. In designing games for the elderly's physical rehabilitation, it is important to take into account that game user interface should be simple and uncluttered. It should provide effective visual cues for elderly so that they can pay more attention to their gameplay. It should design less text-based feedback to elderly. Instead, it should use effective audio and visual feedback to elderly. Gameplay should be simple but effective enough for the elderly to improve their ability in doing exercises. Furthermore, their progress in the game itself should be simple and meaningful for the elderly rather than using points and scores. We also learn that some physical actions are not suitable for the elderly for their safety in playing games (e.g., jump). The controller-free natural movements are suitable for the elderly because its simplicity and ease-of-use.

6 Conclusion

In this study, we conducted the pre-studies of GSH project. We conducted pre-study 1–a literature review on motivational factors for the elderly in rehabilitation. The findings from this literature review highlight insightful game design guidelines how to motivate the elderly. In pre-study 2–a review of commercial games, we reviewed if commercial games provide motivational game designs for the elderly. The findings highlight that the majority of commercial games are not designed for the elderly although they provide some motivational game designs such as multiplayer game and the real-life environment. These findings provide us insightful knowledge of what is required for the elderly in our future game development. In pre-study 3–a usability of Kinect sensor, we evaluated it with elderly participants. The findings show that Kinect has a potential to be used as an input device for the elderly although its built-in gestures are difficult for the elderly. We also learn that it is important to implement elderly-friendly gestures in our future development. In pre-study 4–we evaluated our existing Puuha's game and two commercial games: Xbox Climbing game and PlayStation's Tennis game. The findings from this usability testing show that commercial games are not suitable for the elderly because of their clutter interface, graphics, and gameplay. We also learn that Microsoft Kinect is the most effective input device for the elderly to play a game.

The findings from these pre-studies provide us insightful design guidelines for designing games for the elderly. We also found out the usability of existing commercial games and input devices for the elderly. The usability guidelines from these studies provide us important game design ideas for future game development. Based on these findings, as a future work, we will design and develop a game system for the elderly's physical rehabilitation.

Acknowledgements. This work has been supported by Finnish Funding Agency for Technology and Innovation (Tekes), City of Turku, and several companies. We would like to thank the funding agency, the elderly participants and staffs from the service homes, the game experts from Turku Game Lab, and our industrial partners.

References

1. Al Mahmud, A., Mubin, O., Shahid, S., Martens, J.-B.: Designing social games for children and older adults: two related case studies. Entertain. Comput. **1**(3–4), 147–156 (2010)
2. Boon, S.D., Brussoni, M.J.: Popular images of grandparents: examining young adults' views of their closer grandparents. Pers. Relat. **5**(1), 105–119 (1998)
3. Lloyd, J.: The State of Ingergeneratoinal Relations Today. ILC-UK, London (2008)
4. Ross, N., Hill, M., Sweeting, H., Cunningham-Burley, S.: Grandparents and teen grandchildren: exploring intergenerational relationships. Report for the ESRC, Center for Research on Families and Relationships (2005)
5. Silva, M., Correia, S.: ActiveBrain: online social platform for active and health ageing. Procedia Comput. Sci. **27**, 38–45 (2013)
6. Theng, Y.L., Chua, P.H., Pham, T.P.: Wii as entertainment and socialisation aid for the mental and social health of the elderly. In: Proceedings of 2012 ACM CHI Conference on Human Factors in Computer Systems (2012)
7. Brox, E., Luque, L.F., Evertsen, G.J., Hernández, J.E.G.: Exergames for elderly: social exergames to persuade seniors to increase physical activity. In: The 5th International Conference on Pervasive Computing Technologies for Healthcare (PervasiveHeatlh) (2011)
8. Pyae, A., Tan, B.Y., Gossage, M.: Understanding stroke patients' needs for designing user-centered rehabilitative games. In: The 6th Annual International Conference on Computer Games Multimedia and Allied Technologies CGAT (2013)
9. Marinelli, E.C., Rogers, W.A.: Identifying potential usability challenges for Xbox 360 kinect exergames for older adults. In: The Human Factors and Ergonomics Society Annual Meeting, September 2014, vol. 58, issue 1, pp. 1247–1251 (2014)

10. Kahlbaugha, P.E., Sperandioa, A.J., Ashley, L.: Effects of playing wii on well-being in the elderly: physical activity, loneliness, and mood. Activities Adapt. Ageing **35**(4), 331–344 (2011)
11. Smit, S.T., Sherrington, C., Studenski, S. et al.: A novel dance dance revolution (DDR) system for in-home training of stepping ability: basic parameters of system use by older adults. Br. J. Sports Med. 51(2009)
12. Hansen, ST.: Robot games for elderly. In: The 6th International Conference on Human-Robot Interaction, pp. 413–414 (2011)
13. Pyae, A., Luimula, M., Smed, J.: Understanding motivational factors of stroke patients for digital games for rehabilitation. In: International Conference on Pervasive Games, PERGAMES (2014)
14. Maclean, N., Pound, P., Wolfe, C., Rudd, A.: The concept of patient motivation: a qualitative analysis of stroke professionals' attitudes. Stroke **33**, 444–448 (2002)
15. Krause, N., Frank, J.W., Dasinger, L.K., Sullivan, J.J., Sinclair, S.J.: Determinants of duration of disability and return-to-work after work-related injury and illness: challenges for future research. AMJ Ind. Med. **40**(4), 464–484 (2001)
16. Shimoda, K., Robinson, R.G.: The relationship between social impairment and recovery from stroke. Psychiatry **61**, 101–111 (1998)
17. Sanntus, G.A., Ranzenigo, A., Caregnation, R., Maria, R.I.: Social and family integration of hemiplegic elderly patients 1 year after stroke. Stroke **21**, 1019–1022 (1990)
18. Barry, J.: Patient motivation for rehabilitation. Cleft Palate J. **2**, 62–68 (1965)
19. Finding motivation after stroke or brain damage. http://sueb.hubpages.com/hub/Finding-Motivation-after-Stroke-or-Brain-Damage. Accessed Jan 2016
20. Holmqvish, L.W., Koch, L.: Environmental factors in stroke rehabilitation, being in hospital itself demotivates patients. Br. Med. J. **322**, 1501–1502 (2001)
21. White, G.N., Cordato, D.J., O'Rourke, F., Mendis, R.L., Ghia, D., Chang, D.K.: Validation of the stroke rehabilitation motivation scale: a pilot study. Asisan J. Gerontol. Geriatr. **7**, 80–87 (2012)
22. Flores, E.,Tobon, G., Cavallaro, E., Cavallaro, F.I., Perry, J.C., Keller, T.:. Improving patient motivation in game development for motor deficit rehabilitation. In: Proceedings of International Conference on Advance in Computer Entertainment Technology, pp. 381–384. ACM (2008)
23. Van-Vliet, P.M., Wulf, G.: Extrinsic feedback for motor learning after stroke: what is the evidence? Disabil. Rehabil. **28**, 831–840 (2006)
24. Knight, A.J., Wiese, N.: Therapeutic music and nursing. poststroke rehabilitation. Rehabil. Nurs. **36**(5), 200–215 (2011)
25. Roth, E.A., Wisser, S.: Music therapy: the rhythm of recovery. Case Manag. **15**(3), 52–56 (2004). http://www.xbox.com/en-US/xbox-one/accessories/kinect-for-xbox-one
26. Pisan, Y., Garcia, J.A., Felix Navarro, K.: Improving lives: using microsoft Kinect to predict the loss of balance for elderly users under cognitive load. In: The Interactive Entertainment 2013 (IE 2013) RMIT University, Melbourne, Australia (2013)
27. Webster, D., Celik, O.: Systematic review of kinect applications in elderly care and stroke rehabilitation. J. Neuroeng. Rheabil. **11**, 108 (2014)
28. Ganesan, S., Anthony, L.: Using the Kinect to encourage older adults to exercise: a prototype. In: CHI 2012 Extended Abstracts on Human Factors in Computing Systems, pp. 2297–2302. ACM (2012)
29. Pompeu, J.E., Arduini, L.A., Botelho, A.R., et al.: Feasibility, safety and outcomes of playing Kinect adventures! for people with Parkinson's disease: a pilot study. Physiotherapy **100**(2), 162–168 (2014)

Senior Citizens Experience of Barriers to Information About Healthy Behaviour

Ágústa Pálsdóttir[✉] and Sigríður Björk Einarsdóttir

Department of Information Science, University of Iceland, Reykjavík, Iceland
{agustap,sbe}@hi.is

Abstract. The study examined barriers to information about healthy behaviour, experienced by senior citizens aged 60 years and older. The data was gathered by a questionnaire survey in 2012. Total number of participants was 176. Participants were presented with 13 statements which measure perceived information barriers, a 5-point response scale (1 = Strongly disagree – 5 = Strongly agree) was used. To assess how information barriers relate to age, the participants were divided into two groups, people aged 60 to 67 years and 68 years and older. ANOVA (one-way) was performed to examine difference across the age groups. To examine the effects of sex and education, and how it interacts on the age groups experience of information barriers, factorial analysis of variance (FANOVA) was used. The results suggest that senior citizens are faced with barriers to information that can have impact on their possibilities to promote their knowledge of healthy behaviour. Of the 13 statements, 10 were found to represent information barriers. Sex was found to interact with age for two statements. Education interacted with age for seven statements, with participants with primary education experiencing lower barriers than participants with secondary or university education. Possible explanation to this finding is discussed in the paper.

Keywords: Health literacy · Healthy behaviour · Information barriers · Senior citizens

1 Introduction

The study explores barriers to information about healthy behaviour, experienced by senior citizens in Iceland. The changes in age distribution, with a growing proportion of senior citizens in the world population, poses great challenges. It is expected that from 2013 to 2050 the number of people aged 60 years and older will more than double globally [1]. In Western countries the predictions are slightly lower with the proportion of senior citizens forecasted to double, from 11 % in 2006 to 22 % in 2050 [2].

This means that the welfare society needs to prepare for the increasing number of senior citizens and ensure their prospects for health and wellbeing. An important factor is to encourage them to be actively involved in health promotional interventions through lifelong learning. To obtain the necessary knowledge about healthy lifestyles, they need to have an easy access to quality information that satisfy their needs.

H. Li et al. (Eds.): WIS 2016, CCIS 636, pp. 97–113, 2016.
DOI: 10.1007/978-3-319-44672-1_9

The term health literacy is important in this context. According to World Health Organization [3], health literacy stands for "the cognitive and social skills which determine the motivation and ability of individuals to gain access to, understand and use information in ways which promote and maintain good health (p. 10)." Health literacy, therefore, goes beyond the basic reading or numeracy skills, measured by screening tools such as the Test of Functional Health Literacy in Adults (TOFHLA) [4]. The need for health information to be written so that suits adults with low reading capabilities differs by world regions. The National Institutes of Health in the United States recommends for example that health materials should be written for seventh or eighth grade reading levels, to meet the needs of a typical adult [5]. In comparison, the literacy skills of adults in the Nordic countries are high [6], therefore the necessity for health information to be aimed at individuals with low reading skills is not the same. Health literacy is, furthermore, closely related to a joint definition by UNESCO and IFLA (International Federation of Library and Information Association) of media and information literacy, which allows individuals to "...access, retrieve, understand, evaluate and use, create, as well as share information and media content in all formats..." [7].

The term media and health information literacy is used in this paper, as it combines the concepts of health literacy and media and information literacy. Thus, competency in media and health information literacy is important as a tool for lifelong learning, which provides people with better opportunities to make informed health decisions. Furthermore, there are indications about a positive connection between healthy behaviour and media and health information literacy [8–10]. Various factors, however, may act as barriers that senior citizens perceive as limiting their possibilities to add to their knowledge and understanding of the interrelated aspects of health and lifestyle.

Information barriers have been the subject of a multitude of studies, within information studies as well as other disciplines. A comprehensive overview of the discussion about the issue is not possible within the constrains of this paper. However, several authors have provided literature reviews that may prove useful at shedding a light on the topic [11–14]. The focus here, on the other hand, is on the hindrances that senior citizens experience in relation information about healthy living.

In the past years, the emphasis has been on the digitalization of health information and the restrictions that senior citizens deal with in that respect. This includes for example weak physical condition and health problems which can cause challenges for a certain group of elderly people [15]. Communication barriers, such as problems with the visual and auditory presentation of information [2], and changes in the motor ability which people can experience as they grow older [16] can affect the ability to use digital devices. By taking the needs of older people into account when information technology is designed, for example with suitable interface design and touch screen solutions [17], some of the obstacles that they are faced with might be minimised.

Although barriers related to digital information is an important group of studies, other types of information barriers have been identified, in connection with various social groups. Restrictions in access to information can be caused by a variety of reasons. McKenzie [18] found that people may be reluctant to seek information from health professionals. McKenzie [18] and Dunne [19], furthermore, noted that health professionals are sometimes unwilling, or unable, to provide the needed information. In addition, it has been reported that finding an information source, as well as knowing what kind of information is to be found in it, can be problematic [19, 20]. Cognitive emotional processes are also worth considering, such as emotional distress linked to information or to the information seeking procedure. The tendency to avoid information rather than to seek it has for example been documented in relation to health information [21]. This is, furthermore, related to perceive self-efficacy, where beliefs about poor capabilities can lead to lower motivation to seek information and knowledge [22].

Health information is increasingly being disseminated digitally. Concerns have been raised that because of lack of practice at using the internet and mistrust in the information, senior citizens may not benefit as much from the digitalization as others [23]. Several studies have pointed out that a lack of confidence in an information source can be a hindrance. In particular, health information on the internet may be regarded as less reliable than information from other sources or channels [24–27]. The same applies to beliefs about the lack of utility of information in different kind of sources [19, 26, 28]. Furthermore, previous results have revealed that, although senior citizens sought health information on the internet more frequently in 2012 than they did in 2007, they had at the same time also become more critical of the information and considered it both less useful and less reliable [29]. It is therefore vital to guide them as to where they can access quality health information on the internet. Otherwise, they will be cut off from using it to make rational decisions about their health related behaviour.

Mettlin and Cummings [28] described a number of hindrances that inhibit health communication. They argued that information aimed at promoting a long-term healthy behaviour need to motivate people that may see themselves as healthy, and if information is to be sought out, or even noticed, it needs to be directed at people's interest. They also noted that information that tell people why it is necessary to change their behaviour is not enough, as useful instructions about how to go about it is also needed. Furthermore, they pointed out that information presented through many different channels may result in people receiving conflicting information, and that health information are often considered complex and technical. The last point, difficulties in understanding information has also been found by McKenzie [18].

Apart from research about access and use of digital information, senior citizens perception of barriers to health information has not been widely explored. Nevertheless, there are some indications about hindrances that they face. In a study of senior citizens belonging to the Swedish-speaking minority in Finland, Kristina Eriksson-Backa [30] reported that they experienced barriers such as feeling worried because they did not get enough information about their health care, finding the information contradictory and confusing, and having problems interpreting it. This corresponds with results by Brown and Park [31], who noted that as people grow older, they face more problems in understanding and learning new information. The need for health information that is understandable and can easily be accessed has been stressed. In particular for those groups of older people whose position is not strong, such as people with poor health, lower educational level, and people who are not interested in health information [32].

1.1 Aim and Research Questions

The possibilities of senior citizens to improve their knowledge, in order to make informed choices that promote health and wellness, is a crucial issue which may have impact on the wider prospects for sustainable health and wellbeing. Yet, there is still a number of unanswered questions about the information barriers that they may experience. By identifying these hindrances, the health professionals who are responsible for health promotion are given the opportunity to diminish, or preferably eliminate, them. Subsequently, people's access to health information and their capacity to use it effectively to improve their way of living can be enhanced.

The aim of the current study is to examine the perceived barriers to information about healthy living among people at the age 60 year and older in Iceland. The paper will seek answers to the following questions: (1) What barriers do senior citizens experience in relation to information about healthy living? (2) How do the perceived barriers relate to their age groups, sex and education?

2 Methods

2.1 Data Collection

The data were gathered in spring 2012, using an internet and a telephone survey from two random samples of 600 people each, aged 18 years and older from the whole country. The datasets were merged, allowing answers from all individuals belonging to each set of data. The total response rate was 58.4 %. The current study involves only participants who are 60 years and older. A comparison of the sample and the population shows that there were more people who are 60 years and older in the sample (27.5 %) compared to the population (18.7 %).

In Western countries it has been traditional to use the retirement age to define "elderly" [33], and in Iceland elderly is defined by law as people who have reached the age of 67 [34], when it is usual for people to retire. This has, however, been criticised for not taking into consideration the heterogeneity of senior citizens [35]. It has been pointed out that people's chronological age is less important than determinants, like their physical, cognitive and social capabilities [36]. In accordance with the viewpoints, that there is no clearly defined age when people become senior citizens, the associations for senior citizens in Iceland admit those who have reached the age of 60 to become members [37]. It was decided, in view of this, that people who have reached the age of 60 should be included in the study, and that those who are at the age 60 to 67 years, a group who is approaching retirement, should be compared with people aged 68 years or older, who have reached the retirement age. A total of 176 people participated in the study, 86 women and 90 men. Participants aged 60 to 67 years were 87 while 89 participants were 68 years or older.

2.2 Measurements and Data Analysis

1. Socio-demographic information included traditional background variables. Based on previous analysis the variables sex and education are used in the current study.

Education was measured as the highest level of education completed. Three levels were distinguished: (1) primary education includes those who have finished compulsory education; (2) secondary education includes those who have completed vocational training or secondary school; (3) university education.

2. Age groups. To assess how the experience of information barriers may relate to age, the participants were divided into two groups, those who are aged 60 to 67 years and those who are 68 years and older.

3. Information barriers. Based on the concept of media and health information literacy and the discussion of information barriers above, 13 statements which measure perceived information barriers were developed. A 5-point response scale (1 = Strongly disagree – 5 = Strongly agree) was used for each statement. The statements were categorized in two groups, called physical barriers and cognitive barriers. Physical barriers consisted of three items that refer to the situation that people live in, that is difficulties in getting away from home to seek information, as well as cost hindrances in relation to time and finances. Cognitive barriers consisted of 10 items. Five refer to beliefs about the availability of information, one item refers to the impact of language skills or educational capabilities and four items refer to capabilities of evaluating the relevance of information.

ANOVA (one-way) was performed to examine difference across the age groups for perceived information barriers, for each statement. To examine the effects of sex and education, and how it interacts on the age groups experience of information barriers, factorial analysis of variance (FANOVA) was used.

3   Results

The 13 statements about information barriers were categorized in two groups, called physical information barriers and cognitive information barriers. Results about how each age group experienced the physical information barriers will be are presented first, in Fig. 1. This will be followed by results of how the effects of sex and education interacts on the age groups experience of physical information barriers, where only results about significant differences will be presented.

After that, results about cognitive information barriers will be presented, in the same order as outlined above.

3.1   Perceived Physical Information Barriers

Figure 1 presents the results about the age groups experience of physical information barriers.

The value for all three statements, categorised as physical barriers, are above median (3, Neither agree nor disagree). The statements represent considerable barriers, for both age groups, with values close to or above 4. The values are very similar for the age groups and an examination of each statement, by ANOVA (one-way), revealed that there is no significant difference across them (Fig. 1).

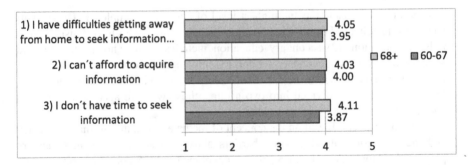

Fig. 1. Physical information barriers

Factorial analysis of variance (FANOVA) was then used to test the statements against education and sex, for each age group. Significant differences were found for statements 1 and 2. The results are presented in Figs. 2, 3, 4 and 5.

Results about statement 1, "I have difficulties getting away from home to seek information about health promotion", in Fig. 2, show a significant difference by age and education. Participants in the younger group with primary education did not agree with this, as opposed to those who have secondary education (p < .000) and university education (p < .000). There was not a significant difference across participants with secondary and university education (p = .853). No significant difference was found by education for the older age group (F(2,160) = 0.2, p = .843).

In addition, significant difference was found by age and sex, with women in the older age group (p < .05) finding it more difficult to get away from home to seek

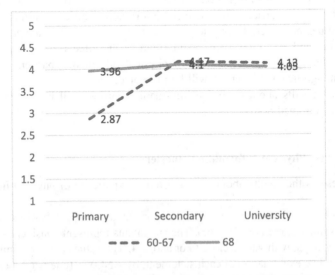

Fig. 2. Difficulties getting away from home to seek information – differences by age and education

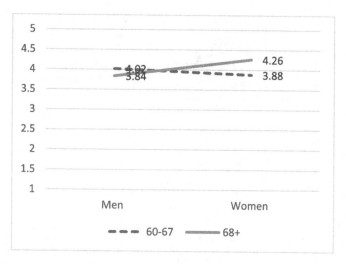

Fig. 3. Difficulties getting away from home to seek information – differences by age and sex

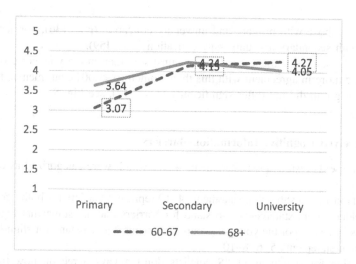

Fig. 4. I can´t afford to acquire information – differences by age and education

information than men, while men and women in the younger age group did not differ significantly (p = .50) (see Fig. 3).

Significant difference was found for both age groups when statement 2, "I can´t afford to acquire information", was tested against education, see Fig. 4. In the younger group, those who have primary education agreed less with this than participants with both secondary (p < .000) and university education (p < .000). Significant difference was not found across participants with secondary and university education (p = .726). In the older group participants with primary education were less in agreement than those who have secondary education (p < .010), while no significant difference was

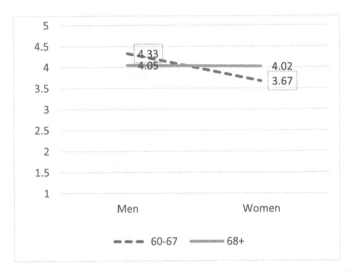

Fig. 5. I can't afford to acquire information – differences by age and sex

found across those with primary and university education (p = .136), nor across participants with secondary and university education (p = .459).

When the same statement was tested against sex, men in the younger group were found to be more in agreement with this than women (p < .001), but men and women in the older group did not differ significantly (p = .912), see Fig. 5.

3.2 Perceived Cognitive Information Barriers

Figure 6 presents results about statements 4–10, which were categorized as cognitive barriers.

Results in Fig. 6 show that statements 4–10 represent barriers for both age groups. For the older group, statements 4–6 stand for barriers that are somewhat higher than statements 7–10 do. For the younger group, statements 4 and 7 stand for slightly higher barriers than statements 5, 6, 8–10.

The value for statement 11 "Specialists don't always agree on how to protect health, therefore I don't know what information can be trusted", statement 12 "The media often publishes information from people whose qualifications I don't know, therefore it's difficult to know what is reliable and quality information", and statement 13 "The amount of information on the internet makes it difficult to choose from", are below 3 (Neither agree nor disagree), indicating that they do stand for barriers.

When each statement was analysed by ANOVA (one-way), no significant differences were found across the age groups. However, when the statements were tested against education and sex for each age group, by factorial analysis of variance (FANOVA), significant differences were found by education for statements 4, 5, 7, 8 and 10. The results are presented in Figs. 7, 8, 9, 10 and 11.

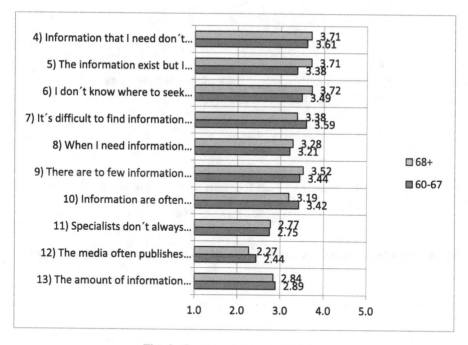

Fig. 6. Cognitive information barriers

Results about statement 4, "Information that I need don't exist", show a significant difference by age and education. Participants with primary education, in both the younger (p < .010) and the older group (p < .05), were less in agreement with this than participants with secondary education. Participants with secondary and university

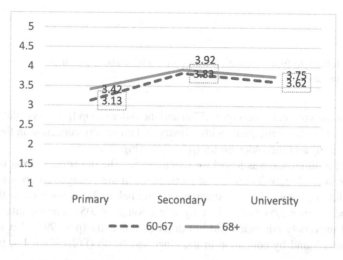

Fig. 7. Information that I need don't exist – difference by age and education

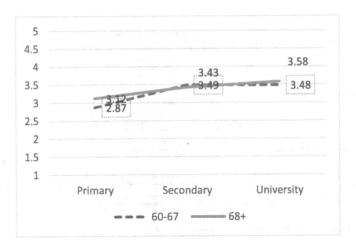

Fig. 8. The information exist but I don't have access to it – difference by age and education

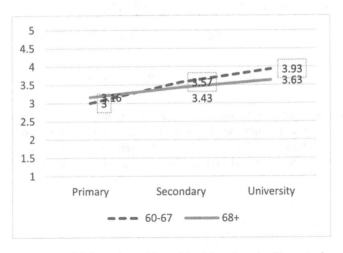

Fig. 9. It's difficult to find information with useful advice about health protection – difference by age and education

education in the younger group (p = .379) and the older group (p = .487) did not differ significantly. Nor did participants with primary and university education in the younger group (p = .086) and the older group (p = .246) (Fig. 7).

Significant difference was found for statement 5, "The information exist but I don't have access to it". Figure 8 shows that participants in the younger group with primary education did not agree with this statement (value below 3), as opposed to those who have secondary (p < .05) and university education (p < .05). Participants with secondary and university education did not differ significantly (p = .994). No significant difference was found by education in the older age group (F(2,162) = 1.4, p = .249).

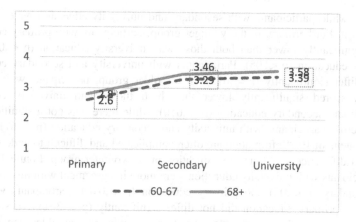

Fig. 10. When I need information about specific items it can be difficult to find – difference by age and education

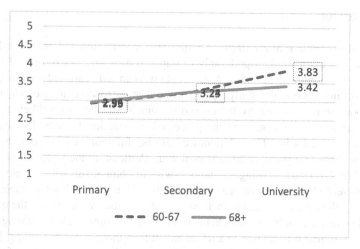

Fig. 11. Information are often complicated and difficult to understand – difference by age and education

Results in Fig. 9 about statement 7, "It's difficult to find information with useful advice about health protection", show a significant difference by education in the younger group. Participants with university degree agreed more with this than those who have primary education (p < .010), while there was not a significant difference across those with university and secondary education (p < .152) nor across participants with secondary and primary education (p < .076). No significant difference was detected by education in the older age group $F(2,159) = 1.2$, p = .298.

Values for statement 8, "When I need information about specific items it can be difficult to find", were below 3 (Neither agree nor disagree) for participants with primary education in both age groups (Fig. 10). They therefore disagreed with this statement, while participants with secondary and university education were in agreement with it. Furthermore, in the younger group, participants with primary education scored significantly lower than both those with university education (p < .010) and secondary education (p < .05). Participants with university and secondary education did not differ significantly (p = .675). In the older group, participants with primary education scored significantly lower than both those with university education (p < .010) and secondary education (p < .010), while there was not a significant difference across participants with university and secondary education (p = .671).

For statement 10, "Information are often complicated and difficult to understand", a significant difference was found by education in the younger group. Figure 11 shows that participants with university education were more in agreement with this than those with secondary (p < .05) and primary education (p < .010). Participants with secondary and primary education did not differ significantly (p = .361). No significant difference was revealed across the educational groups in the older age group F (2,160) = 4.7, p = .328).

4   Discussion

It is crucial to enable senior citizens to acquire the information and understanding that is necessary to maintain, and preferably to improve their knowledge of healthy behaviour. By presenting findings about how people who are 60 years and older experience barriers to information, the study sought to contribute to current research of senior citizens possibilities for healthy living, and sustainable health and wellbeing.

The current study investigated the perceived information barriers that senior citizens are confronted with. These hindrances can be different, at least some aspects of them, from those that younger citizens deal with. Particularly the youngest generation, who is often described as being technology-wise in comparison with the elderly generation. Nevertheless, the results show that senior citizens did not find it problematic to select from information on the internet. Nor did they consider identifying the quality of information in other information channels to be a hindrance. Thus, although senior citizens seek digital information to a lesser extent than those who are younger [38–40], they seem to be confident about their capabilities at critically evaluate the information.

The study results provide some insight to the hindrances that senior citizens are confronted with. Of the 13 statements presented in the study, 10 were found to represent information barriers. The participants' experience of hindrances varied though. On the scale of 1 to 5, the values for the three statements categorised as physical barriers were either above or close to 4, which indicates that the circumstances that senior citizens live in may create considerable hindrances to information. Seven of the 10 statements which were classified as cognitive barriers stood for information barriers. The values for these statements were though more diverse and somewhat lower than for physical barriers. But taken together, these results suggest that senior citizens are faced with obstacles that can have impact on their possibilities to promote their knowledge of healthy behaviour.

No significant difference was found when the age groups were compared. However, a closer examination, where the age group were tested against education and sex, revealed certain trends about the senior citizens experience of information barriers. Significant differences were found more often for participants aged 60 to 67 (statements 1, 2, 4, 5, 6, 8 and 10), than those who are 68 years or older (statements 1, 2, 4, 8). It is not possible to draw broad conclusions from this but the question comes up why this is so.

For only two of the statements, sex was found to interact with age. Women aged 68 years and older found it more difficult to get away from home to seek information than men did, while men aged 60 to 67 years were more likely to feel that they could not afford to get information, than women. Education was found to interact with age for seven of the statements (number 1, 2, 4, 5, 6, 8, 10). The main result here is that for all of these statements, participants with primary education experienced lower information barriers than participants with either secondary or university education, and in some cases both.

At first sight, this finding may appear contradictory, as it seems more logical for those who have higher education to experience lower information barriers than people who are less educated. The key elements of media and health information literacy is that people possess the motivation and the personal skills that allow them to acquire information about healthy living and draw knowledge from it, for their own advantage [3, 5]. Previous research have reported that people who are more educated are also more likely to engage in health information seeking than those who are less educated [9, 41–43]. In addition, the need to pay more attention to motivation, especially how interest in a topic may act as a driving force that inspires people to seek health information, has been stressed in several studies [11, 12, 34, 44, 45]. Thus, more educated senior citizens might also be more motivated and likely to seek information about healthy living and, as a result of this, it is possible that they are more aware of information barriers than those who are less educated.

Although this is merely speculations that serve to seek an explanation, the findings of the study are interesting and give reasons to look further into the connection between the perceived information barriers and the skills of media and health information literacy that people possess. The study results from statements about the availability of information indicate that senior citizens have difficulties seeking information. Therefore, it is suggested that future research could, in particular, pay attention to their ability in this respect and how it can be strengthened. Social cognitive theory and its main component, perceived self-efficacy, might prove useful and come up with suggestions about how such initiative can be improved. Self-efficacy refers to people's expectations about whether or not they will be able to master a certain behaviour, and how successful they will be. Those who are high on self-efficacy beliefs are considered likely to be more strongly motivated, to set themselves higher goals and to have the strength to carry out the act, than those who are low on self-efficacy [22]. Hence, by evaluating the capabilities of senior citizens, in order to discover specific aspects where skills development is needed, the chances of ensuring that they receive appropriate instructions that are aimed at raising both their self-efficacy beliefs and level of competence at information seeking may be improved. Moreover, as interest in healthy behaviour has been recognized as an important motivational factor [28, 32], more research is needed to investigate how it can be enhanced, so that senior citizens may be inspired to practice lifelong learning about healthy living.

Today's information environment is constantly changing. In particular, information and communication technology has rapidly transformed the possibilities to disseminate and access information, a progress that can be expected to continue in the coming years. Hence, in future the questionnaire about barriers needs to be developed accordingly, so that it reflects the various aspects of media and information literacy that people need to be in command of.

The overall study is limited by a rather low response rate of 58.4 %. This is considered satisfactory in a survey bit nevertheless raises the question, whether or not those who answered the survey are giving a biased picture of those who didn't respond. However, the rate of people at the age of 60 years and older in the sample (27.5 %) is higher than in the population (18.7 %), which strengthens the findings. Thus, the study results may provide valuable information about the information barriers experienced by senior citizens.

5 Conclusion

As a key to senior citizens' health and wellbeing, it is important to promote their possibilities to be actively involved in health promotional interventions, through life-long learning. This cannot be achieved while they are confronted with hindrances to information. The results of the study indicate that senior citizens are faced with barriers to information that can have impact on their options to enhance their knowledge of healthy behaviour. The findings shed light on these hindrances, as well as calling for answers to new questions.

Taken together, the results suggest that it is problematic for senior citizens to gain information because of difficulties at getting away from home and cost hindrances in relation to time and finances. In addition, knowing where information are to be found, and the ability to seek it, is challenging for them. It is, therefore, of great significance for those who are responsible for health promotional activities to recognize and address these hindrances, and thereby improve the changes of providing senior citizens with information. In doing so, it is vital, not only to ensure them access to quality information, but also guide them as to where and how they can seek it. The policy implications of the findings are that health authorities and professionals need to work together to find ways to make available information about healthy living that can be easily reached by senior citizens, preferably for free. Otherwise, senior citizens may become a disadvantage group in society.

References

1. United Nations: World Population Aging (2013). http://www.un.org/en/development/desa/population/publications/pdf/ageing/WorldPopulationAgeing2013.pdf
2. World Health Organization: Global Age Friendly Cities: a Guide. WHO, Geneva (2007). http://www.who.int/ageing/publications/Global_age_friendly_cities_Guide_English.pdf
3. World Health Organization: Health promotion glossary. Geneva, World Health Organization (1998)

4. Parker, R., Baker, D.W., Williams, M.V., Nurss, J.R.: The test of functional health literacy in adults: a new instrument for measuring patients' literacy skills. J. Gen. Intern. Med. **10**(10), 537–541 (1995)
5. National Institutes of Health: How to Write Easy to Read Health Materials (2016). https://www.nlm.nih.gov/medlineplus/etr.html
6. Nordic Councel of Ministers: Adult Skills in the Nordic Region: Key Information-Processing Skills Among Adults in the Nordic Region (2015). http://www.keepeek.com/Digital-Asset-Management/oecd/education/adult-skills-in-the-nordic-region_tn2015-535#page62
7. UNESCO: Media and Information Literacy (2014). http://www.uis.unesco.org/Communication/Pages/information-literacy.aspx
8. Enwald, H., Hirvonen, N., Huotari, M.-L., Korpelainen, R., Pyky, R., Savolainen, M., Salonurmi, T., Keränen, A.-M., Jokelainen, T., Niemelä, R.: Everyday health information literacy among young men compared with adults with high risk for metabolic syndrome: a cross-sectional population-based study. J. Inf. Sci. **42**(3), 344–355 (2016)
9. Eriksson-Backa, K.: In Sickness and in Health: How Information and Knowledge Are Related to Health behaviour. bo Akademis Förlag - bo Akademi University Press, bo (2003)
10. Pálsdóttir, Á.: Information behaviour, health self-efficacy beliefs and health behaviour in Icelanders' everyday life. Inf. Res. **13**(1), paper 334 (2008). http://InformationR.net/ir/13-1/paper334.html
11. Gaziano, C.: Forecast 2000: widening knowledge gaps. Journal. Mass Commun. Q. **74**(2), 237–264 (1997)
12. Savolainen, R.: Cognitive barriers to information seeking: a conceptual analysis. J. Inf. Sci. **41**(5), 613–623 (2015)
13. Wilson, T.D.: On user studies and information needs. J. Documentation **37**(1), 3–15 (1981)
14. Wilson, T.D.: Information behaviour: an interdisciplinary perspective. Inf. Process. Manag. **33**(4), 551–572 (1997)
15. Pew Research Centre: Older Adults and Technology Use (2014). http://www.pewinternet.org/2014/04/03/older-adults-and-technology-use/?utm_expid=53098246-2.Lly4CFSVQG2lphsg-KopIg.0
16. Hoogendam, Y.Y., et al.: Older age relates to worsening of fine motor skills: a population-based study of middle-aged and elderly persons. Front. Aging Neurosci. **6**, 259 (2014). http://www.ncbi.nlm.nih.gov/pmc/articles/PMC4174769/
17. Piper, A.M., Campbell, R., Hollan, J.D.: Exploring the accessibility and appeal of surface computing for older adult health care support. In: Mynatt, E., Schoner, D., Fitzpatrick, G., Hudson, S., Edwards, K., Rodden, T. (eds.), CHI 2010: Proceedings of the 28th International Conference on Human Factors in Computing Systems, Atlanda, GA, USA, 10–15 April 2010, pp. 907–916. ACM, New York (2010)
18. McKenzie, P.J.: Communcation barriers and information-seeking counterstrategies in accounts of practitioner-patient encounters. Libr. Inf. Sci. Res. **24**, 31–47 (2002)
19. Dunne, J.E.: Information seeking and use by battered women: a "person-in-progressive-situations" approach. Libr. Inf. Sci. Res. **24**(4), 343–355 (2002)
20. Davies, J., et al.: Identifying male college students' perceived health needs, barriers to seeking help, and recommendations to help men adopt healthier lifestyles. J. Am. Coll. Health **48**(6), 250–267 (2000)
21. Case, D.O., Andrews, J.E., Johnson, J.D., Allard, S.L.: Avoiding versus seeking: the relationship of information seeking to avoidance, blunting, coping, dissonance, and related concepts. J. Med. Libr. Assoc. **93**(3), 353–362 (2005)
22. Bandura, A.: Self-Efficacy: The Exercise of Control. W.H. Freeman, New York (1997)

23. Fischer, S.H., David, D., Crotty, B.H., Dierks, M., Safran, C.: Acceptance and use of health information technology by community-dwelling elders. Int. J. Med. Inform. **83**, 624–635 (2014)

24. Bradford, W.H., et al.: Trust and sources of health information the impact of the internet and its implications for health care providers: findings from the first health information national trends survey. JAMA Intern. Med. **165**(22), 2618–2624 (2005)

25. Eriksson-Backa, K.: Finnish 'Silver Surfers' and online health information. In: Eriksson-Backa, K., Luoma, A., Krook, E. (eds.) WIS 2012. CCIS, vol. 313, pp. 138–149. Springer, Heidelberg (2012)

26. Pálsdóttir, Á.: Icelanders' and trust in the internet as a source of health and lifestyle information. Inf. Res. **16**(1), paper 470 (2011). http://InformationR.net/ir/16-1/paper470.html

27. Miller, L.M.S., Bell, R.A.: Online health information seeking: the influence of age, information trustworthiness, and search challenges. J. Aging Health **24**(3), 525–541 (2012)

28. Mettlin, C., Cummings, M.: Communication and behavior change for cancer control. Prog. Clin. Biol. Res. **83**, 135–148 (1982)

29. Pálsdóttir, Á.: Senior citizens, media and information literacy and health information. In: Kurbanoglu, S., Boustany, J., Špiranec, S., Grassian, E., Mizrachi, D., Roy, L. (eds.) ECIL 2015. CCIS, vol. 552, pp. 233–240. Springer, Heidelberg (2015)

30. Eriksson-Backa, K.: Access to health information: perceptions of barriers among elderly in a language minority. Inf. Res. **13**(4), paper 368 (2008). http://InformationR.net/ir/13-4/paper368.html

31. Brown, S.C., Park, D.C.: Roles of age and familiarity in learning health information. Educ. Gerontol. **28**(8), 695–710 (2002)

32. Eriksson-Backa, K., Ek, S., Niemelä, R., Huotari, M.-L.: Health information literacy in everyday life: a study of Finns aged 65–79 years. Health Inform. J. **8**(2), 83–94 (2012)

33. Thane, P.: History and the sociology of ageing. Soc. Hist. Med. **2**(1), 93–96 (1989)

34. Lög um málefni aldraðra nr. 125/1999 [Act on the Affairs of the Elderly nr. 125/1999]

35. Berger, K.S.: The Developing Person Through the Lifespan, 3rd edn. Worth Publishers, New York (1994)

36. Ries, W., Pöthiga, D.: Chronological and biological age. Exp. Gerontol. **19**(3), 211–216 (1984)

37. Landsamband eldri borgara. http://leb.is/. (Association for senior citizens)

38. Fox, S.: Online Health Search 2006: Most Internet Users Start at a Search Engine when Looking for Health Information Online: Very Few Check the Source and Date of the Information they Find (2006). http://www.pewinternet.org/PPF/r/190/report_display.asp

39. Lorence, D., Park, H.: Study of educational disparities and health information seeking behaviour. CyberPsychol. Behav. **10**(1), 149–151 (2007)

40. Niedźwiedzka, B., et al.: Determinants of information behaviour and information literacy related to healthy eating among internet users in five European countries. Inf. Res. **19**(3), paper 633 (2014). http://InformationR.net/ir/19-3/paper633.html

41. O'Keefe, G.J., Boyd, H.H., Brown, M.R.: Who learns preventive health care information from where: cross-channel and repertoire comparisons. Health Commun. **10**(1), 25–36 (1998)

42. Reagan, J.: The "repertoire" of information sources. J. Broadcast. **40**(1), 112–121 (1996)

43. Pálsdóttir, Á.: Information behaviour, health self-efficacy beliefs and health behaviour in Icelanders' everyday life. Inf. Res. **13**(1), paper 334 (2008). http://InformationR.net/ir/13-1/paper334.html

44. Statistics Iceland: Computer and Internet Usage by Individuals 2012. Statistical Series: Tourism, Transport and IT, 97(33) (2012). https://hagstofa.is/lisalib/getfile.aspx?ItemID= 14251

45. Fox, S., Duggan, M.: Information Triage (2013). http://www.pewinternet.org/2013/01/15/information-triage/

Safety at School Context: Making Injuries and Non-events Visible with a Digital Application

Brita Somerkoski[(✉)]

National Institute for Health and Welfare, University of Turku, Turku, Finland
brita.somerkoski@thl.fi

Abstract. Safety and security have for decades remained basic values in the Finnish society. Extreme violence and unintentional injuries at schools have raised the need of more developed measures to analyze the potential risks. The Green Cross application is seen as an example of how to prevent accidents and how to make the non-events visible for the individuals who work at school. The study explores the usability and usefulness of Green Cross injury reporting application. The data is qualitative, based on 10 (n = 10) end-user interviews representing school and day-care staff.

Based on this study the school risks were unpredictable, connected to human factor issues or persons acting against regulations. It looks clear that Green Cross software works quite well for solving physical or structural risks at the school context. However, the software was not very useful when reporting repeatedly happening cases, like aggressive behavior.

Keywords: Non-event · Learning environment · Pedagogics · Injury · Usability · Usefulness

1 Background

In Finland students' right to safety, security and welfare is mandated in the Basic Education Act "A pupil participating in education shall be entitled to a safe learning environment" [1] A pupil's wellbeing concerns everyone working in the school community as well as the authorities responsible for pupil's welfare services. Extreme violence and unintentional injuries at schools have raised the need of more developed measures to analyze the potential risks. At the same time the society is getting rapidly digitalized. This has happened extremely fast in the learning environment at schools and concepts like *smart learning, E-learning* and *virtual classrooms* have been established [2] it has to be noted, that the learning environment is also a work environment for adults such as teachers, school administration as well as cleaning, kitchen or maintenance staff [3], yet an essential part of these activities in school, for instance cleaning, maintenance and food delivery, are outsourced for economic reasons.

Safety and security have remained as basic values for decades in the Finnish society [4] and therefore safety culture should be visible also during the school day. This paper describes a qualitative that aims to investigate the end-users' perceptions and

© Springer International Publishing Switzerland 2016
H. Li et al. (Eds.): WIS 2016, CCIS 636, pp. 114–125, 2016.
DOI: 10.1007/978-3-319-44672-1_10

experiences of the Green Cross application. In this study school safety and security are seen from pedagogic point of view. This point of view includes the structured learning environment, people and practical solutions made at the school as well as the curriculum all of which create a functional context for teachers' actions. In this study the emphasis is put on the structured learning environment, social issues and practical safety solutions. Accident is an event in which a person dies, is severely injured or sustains a less serious injury. The concept contains two components: the event and the injury [5].

In general, school is a safe place for children and adolescents. Despite the injury reductions and safety improvements over the last 20 to 30 years, injury remains a leading cause of death for children and adolescents in Europe. The child and adolescent injury death rates have decreased also in Finland during the last decades, but the figures still remain almost twice as high as rates in the Netherlands, one of the safest countries in Europe (Fig. 1).

Injury is a leading cause of death among children and adolescents aged 0–19 years in and annually about 2800 Finns die accidentally [7], however the most of the children's injuries happen during the leisure time. The most common types of accidents leading to death among children aged less than 15 years are traffic accidents, drownings and other suffocations [8].

To enhance injury prevention, the process that leads to an injury needs to be studied. We need to know exactly where, when and to whom these injuries happen. [9] The recent studies show that neither incidents nor near-miss cases are systematically recorded or monitored at schools. This can be one reason why preventive actions are not carried out precisely. Yet there are various multi-sectoral target programs and

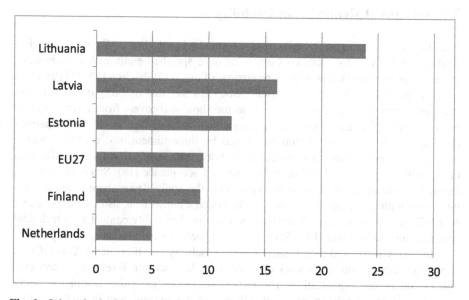

Fig. 1. Injury deaths for children and adolescents, Europe adjusted rate per 100 000 population 0–19 years [6].

action plans for safety and injury prevention in Finland, the most essential of these has been The Internal Security programme that will be replaced with *The Internal Security Strategy* during this year [10]. These strategic documents describe a strong leadership to support the existing infrastructure on children and adolescent safety. More emphasis should be put on implementation of the plans and programs. In these programs, it is recommended to monitor injuries and develop reporting systems for the local needs [11, 12].

On the other hand, accident prevention consists of working towards being accident-free. It is challenging to promote safety, when nothing has happened. Freedom from accident, a non-event, can always be deemed to be a successful end result. Accidents can be prevented, from the top, for instance from an administrative level, or from the bottom, for instance local or individual level. [13] The application presented in this study works both ways, from the school administration to school when analyzing the risks and from the the bottom, school level, when reporting them. However, the focus in this study is at the the school level. For each serious accidental injury there is a number of milder injuries. Only the part of accidents that results in serious physical or material injuries are recorded in the statistics [8]. At the moment there are no nation-wide statistical system that would cover the school injuries and near-miss cases. The Green Cross application that is explored in this study, is seen as an example how to prevent accidents and how to make the non-events visible for the individuals who work at the school for the parents as well as pupils.

The study questions in this study are: How is the Green Cross software used in the pilot schools? What kind of injuries are reported? Is the Green Cross suitable for reporting and analyzing the injuries and near-miss cases in the school context?

2 Conceptual Remarks on Usability

Usability is introduced in the ISO 9241-11 standard as follows: "Extent to which a product can be used by specified users to achieve specified goals with effectiveness, efficiency and satisfaction in a specified context of use" [14]. Nielsen [15, 16] describes the concept of usability with term usefulness. He states that usability of the software consists of how efficient the software is to use, how it recovers from errors, enjoyability, visually pleasant dimensions, memorability and satisfaction [15, 16]. Sharples [17] suggest that usability should be studied by three dimensions: usability (will it work); effectiveness (does it enhance the activity) and satisfaction (is it liked). Also there is often a strong link between usability and acceptance [18]. Since this study is focused on social rather than technical factors, the study design here contains user interviews with a concept of usefulness. The concept means hear also user experience; possibilities and weaknesses. Therefore the traditional usability content with technical specification like mentioned in ISO 9241-11 or Nielsen's definitions are partly faded.

This study is focused in the usefulness and usability questions of the Green Cross application that is web-based work-flow software designed for developing safety culture in educational organization. The project started by designing a tool for safety promotion, problem solving, practical actions and risk management in such way that the safety and risk information could be visually shared at the unit. The software was designed in co-operation with school authorities as a part of regional quality assurance work in five communities. Principal and school administration is in essential role when bringing the safety culture in practice. Safety culture and safety measures at school lie deeply on principal's shoulders. In the schools involved in this study the decision of using Green Cross software was done at municipality level and the end-user teachers could not make any choice whether to use the application or not.

3   Description of Green Cross Risk Reporting Application

Green Cross visualizes the incidents of one calendar month in an easily interpretable format. The screen indicates one calendar month at a time divided into 30/31 units (days). This view is made available to all users so that the whole community can easily see the safety situation in one view. If no incidents have happened, the units in Green Cross remain green. When an incident has occurred and is reported, the units change color according to the classification of the incident. The colour will turn red if the reported case is an actualized event such as an injury or accident, or alternatively yellow in a near miss case. This color-symbolized visual form provides a picture of the safety situation in one glimpse (Fig. 2).

During the cause and risk analysis phase the working methods, people, machines and other physical environment, material and knowledge matters are discussed and

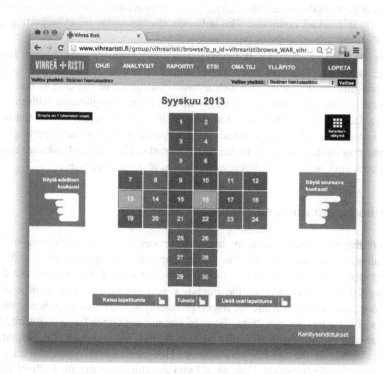

Fig. 2. Screenshot of the basic Green Cross screen

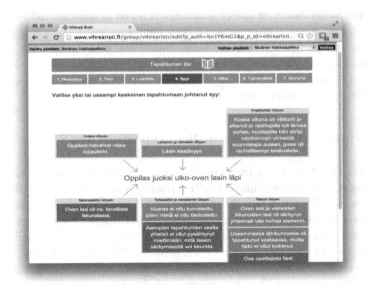

Fig. 3. Screenshot for cause and risk analysis of the Green Cross tool

analyzed in order to understand how the event happened (Fig. 3). Once the reported incident has been analyzed and the agreed safety improvement measures implemented, the analysis is marked complete. The software also provides injury reporting capabilities of all the school units in one municipality.

There are three basic phases in Green Cross safety improvement process: incident reporting; cause analyses and problem solving. The quick incident reporting phase takes approximately 2–3 min, in which a basic description of the case is noted. This paper discusses the usefulness and usability of Green Cross application with the aid of qualitative data. Firstly, it explores what kinds of injuries are reported with the help of the application, secondly the practical actions carried out after the reporting in the monitoring process and thirdly how the product works. The reported risks are dangers, injuries, accidents, violence, bullying or problems at work processes.

4   Methods, Sample and Study Questions

The aim of this study was to examine user experiences, views and definitions of Green Cross software end-users. Since the main approach was qualitative, a structured open-response, semi-structured thematic telephone interviews were carried out during the spring 2015. The content of the interview was based on the idea of usability, effectiveness and satisfaction and the interview questions were based on these concepts. Examples of the interview questions: How do you use the Green Cross software? What is a typical case reported with this software? (usability); Do you think this program can be used to eliminate the risks at your school? How well the program is working for your purposes? (effectiveness); How would you develop the software further? Do you think you could use another method to report and analyze the risks in your school? (satisfaction).

The data were transcribed and analyzed with qualitative content analysis. The study used grounded theory to describe the end-users' perceptions on data usefulness and usability. Two specific actions were completed to get the preliminary conception of the software Green Cross. Firstly, to formulate the semi-structured interviews, a preliminary interview was carried out. The key informant, preliminary interviewed, was a person who is specialized for the design of the Green Cross software. This interview was done to get the preliminary conception of the software in this study. Secondly, the researcher, who was involved with the interviews, learned how to use the software Green Cross.

The sample consisted of 10 (n = 10) software end-users who had been using the Green Cross software two to three years for reporting risks at school. Persons interviewed included teachers, principals, as well as preschool and school administration who worked at the comprehensive school in Pirkanmaa district, which had piloted the use of Green Cross software. The persons that had been involved with the use of Green Cross were chosen as key informants in this study. The average age of the responders was 51.8 years, average working years within education being 22. The average amount of people at the school where responders were working was 337. Of the schools involved in this study 2 were elementary, 6 upper secondary, 1 kindergarten. One of the responders represented school authority.

5 Results

Firstly the study discussed about the usability as a process. All the responders had participated in an education event for 60–120 min for using the software in the beginning of the school year. Some of them were offered additional lessons on a yearly basis. The basic idea of the process was quite similar in all the communities participating in this study.

The reported risks were processed in weekly teacher meetings or additional risk group meetings where the aim was to find out why the risk situation happened and how it could be prevented. A typical case was an injury, accident or an act of violence with/involving a human factor (Table 1).

Totally 38 risk cases (n = 38) were mentioned at the interview of 10 respondents. 21 % (8) of them were near-miss cases. The most typical case that was reported was a failure or malfunctioning of the school property or product.

Examples of the cases reported with Green Cross tool are here classified in five groups (Table 2). The classification is based on a modification of the injury reporting system of the Finnish Rescue Services (PRONTO) [19].

The responders reported that the most common case was a violence based event caused by students, such as aggressive pushing, fight or carrying a knife at school. In the injuries class, the winter time injuries are typical, and also the unpredictable happenings with a human factor. It seems that most of the cases reported are physical or visible.

Table 1. Injuries, accident of near-miss cases at the responder interview

Injury, accident or near-miss	f	%
Trips, falls, risky behaviour	4	11
Slips	4	11
Violence, aggressive behaviour	5	13
Unsuitable object	1	3
Structure, property	10	29
Illness	6	16
Traffic	3	8
Other	4	11

Interview number 1 (I1): "A typical case is when two students start to hit each other. Also during the sports lessons injuries are typical. - -Since there is very limited space at the corridors, in the analyzing session we started to find solutions for how we could have less students going outside at the same time."

Also the practical actions were discussed during the interview. According to the responders it was important to continue the analyzing process after the reporting phase. Basically, the analysis consists of the question; "Why did this happen and how could we avoid this next time?" One of the strengths of the program was the visibility and clearness – the overall safety situation could be seen in one glance. The responders reported about the practical solutions done after Green Cross analyze discussion (Table 3). Also it was stated that Green Cross was the first program where near miss cases were systematically collected.

I1: "When the risk situations are collected and reported, everybody gets to know about what happened. It is important to continue the process to the analyzing phase, if that doesn't happen, it is just another program."
I6: "Prior to this program we were using paper forms to report the risk cases and near misses. It is not as easy to get the whole picture of the situation by reading through the papers. Also it is possible to add pictures to the reports and that is very useful."

Table 2. The reported Green Cross injuries, accidents and near miss cases

Violence: violent behavior, a knife found in student´s clothing, bullying, student throwing objects, other aggressive behavior, a student escapes from the school,
Injuries: icy or slippery surface, a head hit to a stone wall, student fell down at a playground, finger injured by door, teacher was hit by hard baseball, student ran through window glass, student jumped down from storage building roof, allergic reaction, student´s head got stuck between the wall and the staircase
Structural or technical failures: bad acoustics, broken handrail, school door was open, cleaner´s school keys were stolen, electrical appliance was broken, loose object in the door, indoor air pollution issues
Accidents: car accident, student´s work jacket caught fire during crafts lesson
Near miss: student was about to get injured in the angle grind machinery, allergic child got wrong food, vehicle was speeding at the school yard

Table 3. Cases and solutions in a Green Cross analyze phase

Case or near miss	Solutions after Green Cross analyze
Fights in crowded corridors when going to recess	Flexible time table, re-scheduling rush times
Student climbed and jumped off the roof during the recess – severe injury	Supervision changes during recesses, CPR training for the supervisors, modifications to roof access
Near miss: an allergic person got wrong food	Additional education material for substitute workers, better process description for the food delivery staff
Student was stuck at the staircase	Technical department was contacted, additional installations to the staircase
Student's unpredictable aggressive behaviour	Code word for announcing the staff to respond
Work jacket got on fire during craft, design and technology lesson	New and fire-proof working jackets for students
Child ran through the window	More durable window with markings installed
Near miss: a student was about to hurt his hand when using a lathe (turning machine)	New method of working with machinery introduced

The responders reported a few challenges using Green Cross reporting program. Remembering the password when starting to use the program seemed challenging. This was difficult, because the responders were not using the program on a daily basis. It was not clear for the responders who would need the reports and whether the information was needed outside the school or not. Some considered the software as double reporting; it was suggested that there would be a possibility to print forms in the precise format required by the insurance company.

I2: "A grey panther – as an old worker it is difficult to remember all the passwords when you don't use the software more than a couple of times a year."

I7: "- - We can learn from each other and look at the reasons without blaming anyone. Still some of the workers here think that this process is not necessary and we also have workers who are not so used to use computers this way. But I think by reporting the cases we have a possibility to prevent the accidents and injuries."

For some schools and some users, on-line reporting felt difficult. They hoped that Green Cross solution would also classify the risks. This was reported to make the process easier. Also when talking about children with special needs there are continuous risk situations during the day and the users hoped there would be a possibility to combine the cases in a way or another. Despite of these weaknesses all the users believed that the software was helping them to raise and enhance the safety culture at the school.

I7: This program wakes you up, otherwise my table would be filled with little notes and part of them are forgotten forever. This program keeps me in the map, what's going on in the school.

One of the issues taken up by the responders was compatibility. The program seemed to work well when reporting the near-miss cases. However, the compatibility was considered poor. After an injury, teacher would first write a report for Green Cross, in addition for school administration program Wilma and thirdly write a report for the insurance company. Responders suggested possibilities for developing the application further. Some of them also wished Green Cross could be used with a mobile phone.

I5: It would be best if this application could be connected to the school administration Wilma program to avoid double or triple reporting. It is also difficult to add any pictures to Wilma. Green Cross gives a holistic picture of the safety situation and the information can be disseminated to everybody who works at the school.

I6: Pretty often it is the same student or situation that causes all the events. I really hope it would be possible to connect these cases. Otherwise, when reporting, we have to start from the beginning every time.

6 Discussion

This study examined a novel application of web-based technology to enhance schools in promoting safety and reporting injuries, accidents and near-miss cases; the actions carried out after the risk monitoring process and how the product itself works in reporting injuries, violence and near-miss cases.

Findings presented here indicate that the usability and the effectiveness of the software is fairly good. According to the responders of the study, Green Cross is a well-designed program that has the potential to monitor and analyze the risks that would not be analyzed otherwise. The most typical injury during the school day was around structural issues, for instance broken or malfunctioning property. In addition, many school risks were unpredictable, connected to human factor issues, persons acting against norms and regulations or using structures or products in a way they are not supposed to be used. This makes predicting the risks challenging. Green Cross solution provided equally and efficiently a documentation of the whole safety situation in the learning environment. About one fifth of the reported injuries were near-miss cases. Especially these risks could be reported with the help of Green Cross application. These cases would not be reported with any other process at the school context. If a systematic process for school bullying needs to be established with the Green Cross tool, the issue should be better supervised and mentored.

7 Conclusions

The end-user perceptions on the Green Cross software with three main concepts, usability, satisfaction and effectiveness, were explored. Green Cross provided a roadmap and an analyzing method for monitoring and preventing risks. The responders reported cases after which practical actions were carried out. These actions were, for instance, changes in supervision or technical issues. Practical measures can be seen essential for enhancing the safety culture [20–22].

When developing risk analysis solutions for learning environments, a special attention should be paid to satisfaction and acceptance dimensions. The quick incident reporting system was considered very useful, whereas very basic abilities such as memorizing the password were considered challenging due to the nature of teachers' practical tasks during the day [see also 23]. Also, part of the staff such as cleaning or kitchen personnel are not familiar with computers. Some users hoped that the software would become more accessible, for instance applications for phone or other device and also some responders hoped they could have insurance forms in printable form as a part of reporting Green Cross. Based on this study it looks clear that Green Cross software works quite well for solving physical or structural risks at the school context. Yet the software was not very useful when reporting repeatedly happening cases, like aggressive behavior where no new measures could be taken any more. Making the Green Cross application more visible would probably encourage teachers and other staff to report more often. It is important that the administration gives support to data usability to gain sustainability at the program use. This would make the data more comparable within the community.

The results indicate that there are structural and unpredictable incidents and every-day accidents at school, caused mostly by unpredictable human behaviour. In the school context it is vital to consider also other risks than injuries, such as violence and other mental health issues. By monitoring and analyzing the near miss cases it could be possible to prevent accidents to escalate. However, there is still need for more supervision and encouragement to use the software more actively on an every-day basis. It seems, that the user activity goes down, if the personnel is not encouraged to use the application in a sustainable way. As the personnel gets the image that reporting can make the change in safety culture, it would encourage more people to use the software. When designing such software for a learning environment use, usability should be considered as easily accessible tools would probably be more efficient at schools. As a core conclusion it was found that structural risks, unpredictability and the human factor dominate the risks at school. This sets challenges when monitoring the risks. With the help of the Green Cross tool it is possible to make safety culture more visible. This also enables learning from risks and not by shocks.

Acknowledgements. The author wants to thank The Finnish Fire Protection Fund and University of Turku, Rauma unit, that has funded this study as well as all the school staff in the Pirkanmaa district that participated to the study.

References

1. Basic Education Act. http://www.finlex.fi/en/laki/kaannokset/1998/en19980628.pdf. Accessed 18 Sep 2014
2. Gore, V.: E-learning and use of ICT in virtual class rooms. Int. J. Innov. Knowl. Concepts **2** (1), 12–16 (2016)
3. Occupational Safety and Health Act. https://www.finlex.fi/fi/laki/kaannokset/2002/en20020738.pdf. Accessed 5 May 2015

4. Helkama, K.: Suomalaisten arvot. Mikä meille on oikeasti tärkeää? Suomalaisen kirjallisuuden seura. Meedia Zone, Tallinn (2015)
5. Andersson, R., Menckel, E.: On the prevention of accidents and injuries. Comparative analysis of conceptual frameworks. Accid. Anal. Prev. 27(6), 757–768 (1995)
6. Eurosafe: European Child Safety Alliance (2012). http://www.childsafetyeurope.org/publications/info/child-safety-report-cards-europe-summary-2012.pdf. Accessed 15 Mar 2016
7. Somerkoski, B., Lillsunde, P.: Safe community designation as quality assurance in local security planning. In: Saranto, K., Castrén, M., Kuusela, T., Hyrynsalmi, S., Ojala, S. (eds.) WIS 2014. CCIS, vol. 450, pp. 194–202. Springer, Heidelberg (2014)
8. Impinen, A.: Everyday accidents in statistics. In: Somerkoski, B., Lillsunde, P., Impinen, A. (eds.) A Safer Municipality. The Safe Community Operating Model as a Support for Local Safety Planning, pp. 64–82. Juvenes Print, Tallinn (2014). National Institute for Health and Welfare. Directions
9. Somerkoski, B., Impinen, A.: How to survey and monitor the accident situation at local level. In: Somerkoski, B., Lillsunde, P., Impinen, A. (eds.) A Safer Municipality. The Safe Community Operating Model as a Support for Local Safety Planning, pp. 54–56. Juvenes Print, Tallinn (2014). National Institute for Health and Welfare. Directions
10. STM Target Programme for the Prevention of Home and Leisure Accident Injuries 2014–2020. Juvenes Print, Tampere (2013)
11. Markkula, J., Öörni, E.: Providing a safe environment for our children and youth. The national action plan for injury prevention among children and youth. Reports 27/2009. National Institute for Health and Welfare (THL), Helsinki (2009)
12. Lounamaa, A.: Improving information systems for injury monitoring to support prevention at the local level. Opportunities and obstacles. Doctoral thesis. Juvenes Print, Tampere (2013)
13. Somerkoski, B., Lillsunde, P.: What is safety? In: Somerkoski, B., Lillsunde, P., Impinen, A. (eds.) A Safer Municipality. The Safe Community Operating Model as a Support for Local Safety Planning, pp. 44–53. Juvenes Print, Tallinn (2014). National Institute for Health and Welfare. Directions
14. ISO 9241-11 Standard. http://www.userfocus.co.uk/resources/iso9241/part11.htmlScope. Accessed 5 May 2015
15. Nielsen, J.: Usability Engineering. AP Professional, Boston (1993)
16. Nielsen, J.: Introduction to Usability (2012). http://www.nngroup.com/articles/usability-101-introduction-to-usability. Accessed 5 May 2015
17. Sharples, M.: Methods for evaluating mobile learning. In: Vavoula, G.N., Pachler, N., Kukulska-Hulme, A. (eds.) Researching Mobile Learning: Frameworks, Tools and Research Designs, pp. 17–39. Peter Lang Publishing Group, Oxford (2009)
18. Motta, E., Cattaneo, A., Gurtner, J.: Mobile devices to bridge the gap in VET: ease of use and usefulness as indicators for their acceptance. J. Educ. Train. Stud. 2, 65–179 (2014)
19. PRONTO. The Statistical Data System for Finnish Rescue Services
20. Lindfors, E.: Turvallinen oppimisympäristö, oppilaitoksen turvallisuuskulttuuri ja turvallisuuskasvatus – käsitteellistä pohdintaa ja kehittämishaasteita. In: Lindfors, E. (ed.) Kohti turvallisempaa oppilaitosta! Tampereen yliopisto. Kasvatustieteiden yksikkö, pp. 12–28 (2012)
21. Somerkoski, B.: Turvallisuus ja liikenne. In: Niemi, E. (ed.) Aihekokonaisuuksien tavoitteiden toteutumisen seuranta-arviointi 2010. Koulutuksen seurantaraportit 2012:1 (2012)

22. Somerkoski, B.: Turvallisuus yläkoululaisen kokemana. In: Mäkinen, J. (ed.) Asevelvollisuuden tulevaisuus. Maanpuolustuskorkeakoulu. Johtamisen ja sotilaspedagogiikan laitos. Julkaisusarja 2/2013. Artikkelikokoelmat n:o 9, pp. 133–143 (2013)
23. Somerkoski, B.: Injuries at school: digital application Green Cross as a safety audition tool. In: Proceedings of the 9th International Multi-Conference on Society, Cybernetics and Informatics, IMSCI 2015, 12–15 July 2015, Orlando, Florida, pp. 50–53 (2015)

Participative Game Design in the Zet Project – Engaging the Youth to Enhance Wellbeing

Hanna Tuohimaa[1(✉)], Ville Kankainen[2,3], Tarja Meristö[1],
and Jukka Laitinen[1]

[1] Laurea University of Applied Sciences, Lohja, Finland
{hanna.tuohimaa, tarja.meristo,
jukka.laitinen}@laurea.fi
[2] Meanfish Ltd, Tampere, Finland
ville.kankainen@meanfish.fi
[3] The University of Tampere, Tampere, Finland

Abstract. Engagement in meaningful activities is important in supporting the youth in their wellbeing for today as well as for the future. Serious games and game design provide an interesting field for exploring the opportunities and possible benefits for activating the youth at the same time benefiting from their experiences. In this article, a game design process conducted following the principles of participatory design is presented. The article provides examples of the positive as well as the challenging aspects of engaging the youth.

Keywords: Participative game design · Youth · Engagement · Wellbeing · Serious games

1 Introduction

The Finnish comprehensive school ends generally at the age of 15 or 16. After that, the youth need to decide their secondary education, either aiming at a vocational school or upper secondary school. Both paths also enable tertiary education from 18–19 years onwards. According to EuroStat [1], 10.2 % of the Finnish 15 to 24 year olds were neither in employment nor in education and training in 2014. In 2016, the amount of unemployed 15 to 24 year old youth varies between 46 100 and 70 000 persons of the whole population (7.2 % and 10.9 % respectively) depending on the definition and method of data collection [2, 3].

Since 2013, the youth guarantee has aimed at enhancing access to education and employment for the youth in Finland. Every youth is guaranteed a place to study after leaving comprehensive school and every youth under 25 or under 30 in case of recently graduating is guaranteed a place for work, a work try-out, a study place, a place at a workshop, or rehabilitation placement within three months of unemployment. The youth workshops aim at supporting the youth in life management and working skills while at the same time providing guidance for finding the right path for education or employment [4].

The youth are going through several transitional phases where important choices for the future are made. At 25, the youth may already have entered the labor market,

© Springer International Publishing Switzerland 2016
H. Li et al. (Eds.): WIS 2016, CCIS 636, pp. 126–135, 2016.
DOI: 10.1007/978-3-319-44672-1_11

may have graduated but be without employment, may be in the middle of studies, or may still be in search of the right education for the future. Several possible paths exists. In addition to conscious and goal oriented choices, chance and contingent events shape the path and mingle the everyday life of the youth. For many, youth is a period in life full of exciting opportunities. However, the uncertainty of the future, the multitude of decisions to be made combined with all the new requirements of adulthood may also make youth a demanding period in life causing challenges to the youth's present and future wellbeing.

One of the key questions in society is how to support the youth to find the right path and to feel well today as well as tomorrow. Two central elements of wellbeing are to have meaningful ways to occupy one's time and to be engaged in the community [5]. In order to support the youth, then, it is critical to offer the youth opportunities to engage in activities that make sense to them and have an influence on the surrounding society.

Several studies link self-efficacy beliefs to better motivation in health practices [6, 7]. Self-efficacy is a term of the social cognitive theory that claims that beliefs about the ability to exercise control over one's actions are crucial in attaining goals [8]. Improving people's self-efficacy beliefs and their feeling of control can therefore be seen as enabling movement towards better wellbeing. In the context of youth, positive youth development is one of the schools of thought aiming at developing settings for the youth to support initiative by engaging the youth for goal-directed efforts fostering intrinsic motivation [9].

To engage the youth, serious games offer many opportunities. An extensive literary review by Connolly et al. [10] suggests that serious games and videogames have potentially positive impacts on many different areas, particularly focusing on health, business and social issues. Most common benefits were seen on the fields of knowledge acquisition/content understanding, affectivity and motivation. The study further suggests that players find learning by playing both motivating and enjoyable.

In addition to playing, also game design has potential for engaging the youth in a positive manner. Participatory design (PD) is at the same time a design approach, and a design research method, where the users partake in the design of a product. The underlying motivation is to democratize the design process. The reasoning behind this lies on the idea that the people who use the end-product know best how it should work, and thus should have a say in the design process. It is also seen as empowering for the users to take their knowledge as a part of the design process [11].

To some extent, participatory design has also been used in game design - especially while designing serious games [12, 13]. While the game designers are usually adults, the value of this method lies in the collaborative nature that allows the youth to take part in the design process. Lochrie et al. [13] for example used PD in a game design project aimed to connect digitally excluded 11–19 years old people. They argue that it is vital for the success of this kind of project to build a relationship of trust with the participants, and to keep more direct communication between developers and the community. According to Khaled and Vasalou [12] the problem with PD in serious game design is that the users must be fluent with both the game design and the subject matter. As such they suggest confining the workshop content into a subject that relates the expertise of the participants.

When working with people with varying backgrounds and sources of interests, it can be challenging to engage them in active participation [14]. This is especially the case with young people. While observing 13–15 year old teenagers in a PD workshops, Iversen et al. [15] found that teens' motivation was drawn from different kinds of rewards, and while offering refreshments kept them engaged only a brief moment, endorsements sustained their motivation for a longer time. Further, the effect of tools used to engage teenagers in the PD process was seen as highly dependent on how well they are appropriated and valued in relations between the teenagers.

This article is a practice oriented description of a participative game design process in the Zet project developing a game with the youth, for the youth. The goal of the project was to implement participatory design for developing a game that the youth could relate to for managing their everyday life and the future in a fun way. We wanted to explore what the most pressing issues for the youth would be and how they could be transformed into a game concept providing both serious content as well as entertainment. Engaging to youth as experts in their own life in game design is expected to produce a game with high potential for addressing the youth. The game can then be used as a platform for interaction with the youth in youth services, at school or where ever serious subject matters are discussed with the youth about the youth's life. As an end product, the project will also produce a guide book for participative game design to be used in other settings.

The focus in this article is on the process, how the youth have been engaged and what kind of benefits are expected as byproducts of the actual game development goals of the project. The project also provides an example of how the youth may be engaged working as experts in the development of a serious game for their peers.

2 Zet as an Ongoing Case Example

Zet project is a collaborative project funded by the European Social Fund with a game company Meanfish Ltd responsible for the game design and the higher education institution Laurea UAS responsible for reaching the youth and collaborating with local partners in the region In the Zet project, young people develop in collaboration with game design professionals a game for the youth for visioning their future and planning and managing their life. The goal is especially to reach the youth pondering on their future educational choices or career choices between 15 to 24 years of age. However, the project follows an inclusive project design, where all young people are welcome to participate.

The goal of the game design process is to benefit from the experiences and knowhow of the youth in order to design a game which suits the concrete needs the youth have in their everyday life and which speaks to the youth in their own language. At the same time, the process gives the youth success experiences and opportunities to be engaged and have an influence in society. The design process is carried out with two teams meeting in Hanko and Lohja in the Uusimaa region in southern Finland. Lohja is the larger of the towns with 47 300 inhabitants while in Hanko the population is 8900 [16]. The distance between the towns is about 80 km. The youth services in both towns are key partners in the project to reach the youth. In Lohja, the meetings are organized in the premises of Laurea UAS and in Hanko in a local workshops for the youth.

The Zet project also gathers local actors together in network workshops. Service concepts are developed and piloted to support the use of the game. Based on the experiences in the Zet project, a guide book of the participative design process will also be formulated.

As a result of the project, an open source game is designed to be used by the youth either independently or as part of other services, as designed in the project. Concrete benefits are expected also for the participating youth. Engagement, the opportunity to make use of one's know how and experiences and to be active and gain experience in an interesting topic is expected to have positive effects on the self-efficacy beliefs and feeling of control of the youth already during the project. To assess the effects of the design process, a survey is conducted at the beginning and the end of the process. Also interviews will be conducted to gather qualitative data of the design process and its effects.

2.1 Background Information Prior the Game Design Process

Prior to starting the actual game design process, 67 interviews among young people mostly in the Lohja region were conducted in September–October 2015 by nursing and business students at Laurea UAS. The questionnaire was designed by Laurea personnel representing futures research, business and nursing. The goal was to gather background information of the youths' future visions within 5–10 years and present day health and wellbeing in the region. The sub-themes of the desired futures were related to housing, studying, work, voluntary work, friends, family and hobbies. Also background information from the participants were collected. The age of the interviewed people varied from 15 to 24, average being around 20 years.

The future dreams of the youth were quite ordinary. Most wanted to live in their own apartment and only a few wanted to move abroad. The most wanted to study in the future. Some of the youth had certain dream professions in mind while others had several options or had no idea at all. Generally, meaningfulness was seen the most essential feature of work. Only few were involved in voluntarily work. The friends of the respondents were mostly from the immediate surroundings even though some had friends from virtual environments, too. The family dreams were quite traditional: the most wanted to have a family with children or at least find a partner. The most of the respondents had some kind of hobbies, often related to sports. The time scale with which the respondents considered the future was maximum one year onwards.

Basically, the respondents found themselves healthy and happy. Over 95 % found their health situation to be very good, good or quite good. Only three persons perceived their health to be rather bad. Almost 90 % were happy or very happy with their lives in general. Seven respondents found themselves quite unhappy or unhappy. The most common health and wellbeing problems were neck and shoulder pains, tiredness, stress, difficulties to wake up in mornings and headache.

After the interviews, a future workshop was organized in October to the youth on the themes of studying, employment, leisure time, family and friends, health and wellbeing and housing to define future visions. In addition, discussions on issues supporting and hindering wellbeing in the present day were also organized. All the information collected were used as background data in the game design process.

2.2 The Game Ideation

The game design process started with a two day game jam resembling ideation session in November 2015. Game jams take a variety of different forms, but a recent definition by Kultima [17] states that: "A game jam is an accelerated opportunistic game creation event where a game is created in a relatively short timeframe exploring given design constraint(s) and end results are shared publically." In the recent years, the popularity of game jams has been on a constant growth, and every year a vast number of jams are organized around the world, varying in size from The Global Game Jam, which in 2015 attracted 28 837 participants around the globe, to small events with only a handful of attendees [17].

Unlike in regular game jams, the goal in this project was to ideate together and not develop full games during the event. The event was advertised vastly in the region the message being that the event is open for everyone without any competence requirements. After coverage on a local newspaper, participants from Lohja started enrolling to the event already in September. Although transportation from Hanko was arranged to the event, youth from Hanko on the other hand was extremely difficult to reach and get interested in participating. Especially the time of the event during the weekend seemed to work against the goal to reach the youth. Eventually, there were 13 youth enrolled to the event with 10 participating. Also one youth worker and one collaborating partner as well as students from Laurea UAS participated in the event.

The game jam event exploited the results of the October discussion events and started generating ideas for the identified challenging situations in the youth's life. The youth listed opportunities and challenges in four themes: What is it like to move into a new city; how to get the dream job; how to be in contact with family and friends and how to keep up one's own wellbeing? After that, by using different ideation and story generation methods, the themes started to convey into ideas for games.

Between Saturday and Sunday, the ideas were themed by the project staff. On Sunday, three of the most interesting ideas were chosen and developed further in teams. One team focused on a game where the player managed an enterprise imperium and sabotaged opponents. In one game, the player fought for his living environment against the bad potato. The third game envisioned the difficulties of interacting with others in case you were a youth struggling with anxiety.

The teams presented their game ideas to the group and the good points of each idea were then discussed together. Turning the roles upside down, humor, an intense atmosphere and well thought game settings were characteristics in the games that were well received. The event was wrapped up by making a combination of the three presentations for further development.

2.3   The Game Theme and Mechanics

In the game, which goes by the working title "Reptilian Overlords", players take the role of a recruiter for a reptilian controlling an international corporation. The aim of the game is to grow your own business, while sabotaging your competitors - AI controlled corporations. The game can be divided into two "modes": controlling your own company, and conducting espionage missions on your competitors' companies.

The approach taken is an upside down view on the challenges of life management and planning for the future (i.e. looking for a job or a place to study). The game concept draws from the benefits of the dark humor created by the upside down situation, as well as distancing the actual game play from the serious content it represents. The aim is to reach an experience where the player does not realize she is learning while playing.

The game mechanics can be divided roughly into three categories by the main aspects of the game: espionage quests, company management and recruiting employees. The espionage quests are conducted with a point-and-click mechanism. These three categories arose from the themes.

On espionage missions, the player controls one of the employees - sent as an agent to a competing company - as the agent tries to steal the blue prints of some device or sabotage the production line. The quest takes the form of a puzzle, where you need to find the right chain of actions that will lead to the desired outcome.

The game characters used in the espionage quests have their distinct characteristics with which the player needs to learn how to cope with and benefit from. All have both positive as well as challenging characteristics. For instance, playing an espionage quest with an employee fearing crowds requires different tactics than playing the quest with a talkative employee. On the other hand, the first may be more attentive to detail while the latter more easily distracted. The right amount of challenge enhances the game characters abilities to overcome difficulties and utilize strengths. The goal is to offer the players insight into how personal differences affect and can be taken advantage of in the working life. Besides the players' own abilities and hindrances, the player may also be able to look at others from a new perspective.

As the game starts, the player as recruiter is unaware of the boss being a reptilian. During the game, the player comes to realize the unethical working methods of the boss and the grievances in the company. With a twist in the game plot, the player then starts to collect evidence against the boss. In the end, the boss is revealed and the player may restore the order in the company. The end twist balances the story arc and the game may be finished with a positive spirit offering many topics for discussion after the game.

2.4   Game Design and Local Team Meetings

The base for the game design team in Lohja was grounded on the participants of the game jam event. A core team of 8 youth in Lohja have participated throughout the project consisting of three youth outside employment or education, one with employment and four students. Youth that have come to the project later or visited the team meetings occasionally, have mainly been students. Altogether 16 people have participated in the game design team in Lohja.

The situation in Hanko was rather different, as we did not manage to engage any youth from Hanko in the game jam event in Lohja. Instead, the first team meeting was the start of the game design process in Hanko. In a larger town, youth interested in participating in game design was easy to find while in a smaller town the aid of the youth workers in engaging the youth was crucial for project success. Through collaborating with a local media workshops and a youth workshop, we were able to reach the core target group of the project i.e. the unemployed youth. However, we did not manage to attract a core team for the game development. We started the team meetings at the media workshop with youth from the youth workshop also visiting the team meetings occasionally. However, the media workshop started cancelling the team meetings due to scheduling issues and the project needed to reorganize its activities. Instead, one game demo presentation and one workshop was organized with the youth at the youth workshop. Therefore, the composition of the team in Hanko has varied considerably throughout the project. Altogether 15 people participated in the game design team meetings in the media workshop and approximately ten attended the events in the youth workshop.

The teams had game design meetings throughout the winter and spring with an interval of 2 to 4 weeks. These meetings were structured into workshops, where the game development experts first introduced the topic of the workshop and then gave a brief presentation on the game development fundamentals. After this, the participants were engaged into rapid game developing exercises. The team meetings also included information snapshots to support game design. For instance, trends and weak signals for the future working life and future professions were presented.

The first workshops focused on iterating the game ideas into a workable game concept. Since the iterative nature of the game design process, the content of the following workshops was kept open during this initial stage. As the game concept started to take a form with solid game mechanics in place, the topics of the remaining workshops was finalized. This gave structure for the entirety of the development process, and assisted communicating the process to the participating youth.

The contents of the workshops during the mid-stage of the development process varied from testing the initial game mechanics with paper prototypes, to designing the character gallery and testing out the digital game development environment to be used in the development. Between the workshops, the new content created and results achieved were implemented on the game. The overall design was thus a collaboration between the experts and the youth participants. The final team meetings will take place in August to assess the game demo before piloting during the Autumn.

2.5   Evaluation of the Process

To gain insight into the expectations of the participants and evaluate the possible benefits of participation, an online starting survey was conducted which will be followed by an ending survey at process end. Follow up interview will also be conducted by a nursing student of Laurea UAS during the autumn. The survey was sent to 15 respondents in Lohja and 12 in Hanko with 15 responses altogether. The response rate in Lohja is 73 % but in Hanko only 25 %. This follows the general pattern in the project of more commitment to the process in Lohja.

The survey focused on the expectations and preferences of the respondents in relation to the game design process. In addition, a ten item general self efficacy scale was included [18] (with a Finnish translation by Härkäpää as described by Aalto [19]). The project timeline did not allow the construction of a specific self-efficacy scale although specificity is acknowledged as an important factor in measuring self-efficacy.

One of the goals of the game design is to make visible to the youth their own capabilities. The target is that at the end of the process the youth have both learned new things as well as learned to capitalize prior capabilities. The 15 respondents so far have expected slightly more to learn new things than capitalize prior capabilities. Items such as familiarity with games, coding, game design, graphics and writing skills are mentioned as prior capabilities. For new capabilities, most are quite general, wishing to learn about all aspects of game development. Also team working skills are often mentioned. The youth list between zero to three capabilities in both questions. In the ending survey, the goal is that the youth will be able to name more capabilities, prior and new, than they anticipated in the beginning.

The self-efficacy scale includes 10 items rated from 1 to 4. The mean of all answers in the starting survey is 3.15. The responses show that the respondents are quite confident on their abilities to solve problems when they try hard enough. Facing opposition or unexpected events, on the other hand, seem to cause more disagreement with the items.

Feedback has also been collected during every team meeting. The questionnaire consists of two sets of closed questions with the first focusing on the content of the event and the second on the conditions of the event. The open questions give the participants the opportunity to make suggestions for game design and future team meetings, list issues that they wish to learn more about and comment all issues freely.

The role of the team meeting feedback questionnaire has mainly been to guide the game design and steer the development process to those themes that especially interest the participants. For instance, the original goal was to get the first digital draft of the game out in January for the youth to evaluate and develop further. As the digital turn was delayed, there was a clear fatigue of ideation and design on paper present in the teams. As the digital platform for game development was presented in March, it provided the game design with a new edge that immediately improved the feedback especially in Lohja.

3 Conclusion

The Zet project is an ongoing case example of participative game design. Therefore, results of its possible benefits to the participants are not yet available. The goal of this article has been to present the framework and design of the project and provide the readers with practical examples of the positive as well as challenging experiences during implementation. Lessons learned may be of use to others interested in supporting the youth in their wellbeing, life management and gaining a future orientation to life through engagement in meaningful activities.

The two game design lines have proven that the sense of meaningfulness is crucial from the point of view of successful engagement. The game design team in Lohja participated in the first ideation sessions in the game jam and the game development followed the game jam smoothly. In Hanko, on the other hand, the commitment to the game concept was weaker. This might originate from the fact of not participating in the game jam, and thus not feeling ownership of the game concept. In addition, organizing the team meetings in a youth workshop made the activity something that the youth were obliged to participate in. Instead, in Lohja the participation was based on voluntariness and interest. Hence, the youth in Lohja were more interested in games per se, and were therefore an easy target group to work with. In Hanko, the interests of the participants varied more. All these three factors made the game design more challenging in Hanko. The planned schedule and work plan needed to be tailored during the process in order to respond to the identified problems.

Although the process in Lohja proceeded more smoothly, from the point of view of developing the participative process, the lessons learned in Hanko were valuable. Had the project been carried out solely in either one of the locations, understanding of the issues affecting the success of the process would have been much weaker. The Hanko case pinpointed the need for meaningfulness and motivation. As the original goal of the project was to offer the opportunity to engage in game design to all the youth irrespective of their personal interest on games, the Hanko experience forced the project group to think through the elements in game design from different perspectives. By tailoring the process, the emphasis may be placed on issues that are of most interest to the participants. On the other hand, dividing the team meetings into sub teams also enables more varied tasks for different interests. As the participants have different kinds of prior knowledge and interest when they come to the project, their roles in the project may also be varied. However, the feeling of equity within the team is also important; all the team member roles need to be equally valuable to the process.

All the tailoring that is required when working with a diverse team makes the management of the design process challenging. The more concrete the tasks of a team meeting are, the easier it is for the youth to participate even without prior knowledge. The process managements' task is then to fix all the pieces together and make sure to keep the big picture of the game design. The team members need to understand where all the tasks are leading, too. Then, interest to learn more on theory, game mechanics and all the serious aspects of the game concept may be awoken as well. It should also be pointed out that although the process in itself is of importance, the end product, i.e. the game and the ways in which the game may be utilized to fulfill its serious functions, is at least as important. The youth do not participate in the game design only to be engaged themselves but instead, they work as experts whose input is crucial for the development of a game with potential to reach and address the youth pondering on their future.

References

1. Eurostat: Young people neither in employment nor in education and training by sex and age (NEET rates) (2016). http://appsso.eurostat.ec.europa.eu/nui/show.do. Accessed 7 Apr 2016
2. Statistics Finland: Labour force survey 2016, February 2016. http://www.stat.fi/til/tyti/2016/02/tyti_2016_02_2016-03-22_en.pdf. Accessed 7 Apr 2016

3. Ministry of Employment and Economy: Employment Bulletin, February 2016. https://www. tem.fi/files/44909/TKAT_Feb_2016.pdf. Accessed 7 Apr 2016
4. Oivo, T. et al.: Youth guarantee 2013 (2012). http://www.tem.fi/files/34025/Social_ guarantee_for_youth_2013.pdf. Accessed 7 Apr 2016
5. Rath, T., Hartner, J.: Wellbeing. The Five Essential Elements. Gallup Press, New York (2010)
6. Marks, R., Allegrante, J.: A review and synthesis of research evidence for self-efficacy-enhancing interventions for reducing chronic disability: implications for health education practice (part II). Health Promot. Pract. 6(2), 148–156 (2005)
7. Strecher, V., DeVellis, B., Becker, M., Rosenstock, I.: The role of self-efficacy in achieving health behavior change. Health Educ. Q. 13(1), 73–91 (1986)
8. Bandura, A.: Health promotion by social cognitive means. Health Educ. Behav. 31(2), 143–164 (2004)
9. Larson, R.: Towards a psychology of positive youth development. Am. Psychol. 55(1), 170–183 (2000)
10. Connolly, T.M., Boyle, E.A., MacArthur, E., Hainey, T., Boyle, J.M.: A systematic literature review of empirical evidence on computer games and serious games. Comput. Educ. 59(2), 661–686 (2012). http://doi.org/10.1016/j.compedu.2012.03.004
11. Spinuzzi, C.: The methodology of participatory design. Tech. Commun. 52(2), 163–174 (2005)
12. Khaled, R., Vasalou, A.: Bridging serious games and participatory design. Int. J. Child-Comput. Interact. 2(2), 93–100 (2014). Special Issue: Learning from Failures in Game Design for Children
13. Lochrie, M.I., Coulton, P., Wilson, A.: Participatory game design to engage a digitally excluded community. In: 5th DiGRA: Think Design Play (2011)
14. Johansson, M., Fröst, P., Brandt, E., Binder, T., Messeter, J.: Partner engaged design: new challenges for workplace design. In: Participator Design Conference PDC 2002, pp. 162–172 (2002)
15. Iversen, O.S., Dindler, C., Hansen, E.I.K.: Understanding teenagers' motivation in participatory design. Int. J. Child-Comput. Interact. 1(3–4), 82–87 (2013)
16. Statistics Finland: Preliminary population by sex and area, February 2016 (2016). http:// pxnet2.stat.fi/PXWeb/pxweb/en/StatFin/StatFin__vrm__vamuu/005_vamuu_tau_101.px/? rxid=6ba6f32c-500b-4f79-8690-2e7aa046bf68. Accessed 7 Apr
17. Kultima, A.: Defining game jam. In: Proceedings of the 9th International Conference on the Foundations of Digital Games, vol. 15 (2015)
18. Schwarzer, R., Jerusalem, M.: Generalized self-efficacy scale. In: Weinman, J., Wright, S., Johnston, M. (eds.) Measures in Health Psychology: A User's Portfolio. Causal and Control Beliefs, pp. 35–37. NFER-NELSON, Windsor (1995)
19. Aalto, P.: Elämähallinnan turvaverkko. Tarkastelussa työssäkäymisen ja parisuhteen merkitykset elämänhallinnalle kuntoutukseen hakeutuneiden keskuudessa. University of Tampere, Tampere (2006)

Analytics Issues of e-Health and Welfare

An Assessment to Toxicological Risk
of Pesticide Exposure

Cristina Coelho[1], M. Rosário Martins[2], Nelson Lima[3],
Henrique Vicente[1,4], and José Neves[4(✉)]

[1] Departamento de Química, Escola de Ciências e Tecnologia,
Universidade de Évora, Évora, Portugal
cristina.argente@gmail.com, hvicente@uevora.pt
[2] Departamento de Química, Laboratório HERCULES,
Escola de Ciências e Tecnologia, Universidade de Évora, Évora, Portugal
mrm@uevora.pt
[3] Centro de Engenharia Biológica, Micoteca da Universidade do Minho,
Universidade do Minho, Braga, Portugal
nelson@ie.uminho.pt
[4] Centro Algoritmi, Universidade do Minho, Braga, Portugal
jneves@di.uminho.pt

Abstract. On the one hand, pesticides may be absorbed into the body orally, dermally, ocularly and by inhalation and the human exposure may be dietary, recreational and/or occupational where toxicity could be acute or chronic. On the other hand, the environmental fate and toxicity of the pesticide is contingent on the physico-chemical characteristics of pesticide, the soil composition and adsorption. Human toxicity is also dependent on the exposure time and individual's susceptibility. Therefore, this work will focus on the development of an *Artificial Intelligence* based diagnosis support system to assess the pesticide toxicological risk to humanoid, built under a formal framework based on *Logic Programming* to knowledge representation and reasoning, complemented with an approach to computing grounded on *Artificial Neural Networks*. The proposed solution is unique in itself, once it caters for the explicit treatment of incomplete, unknown, or even self-contradictory information, either in terms of a qualitative or quantitative setting.

Keywords: Pesticide exposure · Toxicity · Environmental fate · Artificial intelligence · Logic programming · Knowledge representation and reasoning · Artificial neuronal networks · Incomplete information

1 Introduction

Pesticides are extensively used in agriculture aiming at the control of weeds or plant diseases, which may remain in residual amounts in fruits, vegetables, grains, and water, just to name a few. They stand for xenobiotic compounds for living organisms, and their toxicity is not due to a single molecular event or interaction, but rather to a set of occurrences, starting with pesticide exposure and reaching a point of highest development with the expression of one or more toxic endpoints. These happenings include

© Springer International Publishing Switzerland 2016
H. Li et al. (Eds.): WIS 2016, CCIS 636, pp. 139–150, 2016.
DOI: 10.1007/978-3-319-44672-1_12

adsorption, distribution, biotransformation, distribution of metabolites, interaction with cellular macromolecules and excretion [1]. Biotransformation may result in the formation of less toxic and/or more toxic metabolites, while the various other processes determine the balance between toxic and a nontoxic upcoming [2]. Pesticides can be absorbed by oral, dermal, nasal and/or ocular exposure. Human exposure can be dietary recreational and/or occupational, and toxicity could be acute or chronic. Aggregate exposure and risk assessment involve multiple pathways and routes, including the potential for pesticide residues in food and drinking water, as well as in residues from pesticide use in residential and non-occupational environments [3]. To ensure the safety of the food supply for human consumption, *Maximum Contaminant Levels* (*MCLs*) sets the legal limits for the amount of pesticides allowed in food and drinking water. This is related with *Acceptable Daily Intake* (*ADI*), defined as the amount of a chemical that can be consumed safely every day [4].

Pesticide environmental fate and toxicity depends on the physical and chemical characteristics of pesticide, the soil composition, soil adsorption, and pesticide residues found in different soil compartments. The human hazard is determined by the pesticide properties, exposure time and the individual's susceptibility, affecting the magnitude of these processes and the final fate and toxicity of pesticide [5]. Indeed, agricultural pesticides are incorporated into the organism by different routes which can be stored and distributed in different tissues, leading to an internal concentration that can induce alterations, adverse effects and/or diseases. Often, the human exposure to pesticides was evaluated only by human biomonitoring, i.e., measuring levels in matrixes such as blood and urine. However, the concentration measured might not relate to toxic effect [5, 6].

Recent works established pesticide impact and toxicity based on chemical properties, environmental fate and exposure considerations [7–9]. However, those methodologies for problem solving are not able to deal with incomplete data, information or knowledge. Indeed, for the development of intelligent decision support systems aimed at integrated pesticide toxicological risk assessment, it is necessary to consider different conditions with intricate relations among them. Thus, the present work reports the founding of a computational framework that uses knowledge representation and reasoning techniques to set the structure of the information and the associate inference mechanisms, i.e., it will be centered on a *Proof Theoretical* approach to *Logic Programming* (*LP*) [10], complemented with a computational framework based on *Artificial Neural Networks* (*ANNs*) [11].

2 Knowledge Representation and Reasoning

Many approaches to knowledge representation and reasoning have been proposed using the *Logic Programming* (*LP*) epitome, namely in the area of *Model Theory* [12, 13], and *Proof Theory* [10, 14]. In the present work the *Proof Theoretical* approach in terms of an extension to the *LP* language is followed. An *Extended Logic Program* is a finite set of clauses, given in the form:

$\{$

$\quad p \leftarrow p_1, \cdots, p_n, not\ q_1, \cdots, not\ q_m$

$\quad ?\,(p_1, \cdots, p_n, not\ q_1, \cdots, not\ q_m)\ (n, m \geq 0)$

$\quad exception_{p_1}$

$\quad \cdots$

$\quad exception_{p_j}\ (0 \leq j \leq k),\ (being\ k\ and\ integer\ number)$

$\}:: scoring_{value}$

where "?" is a domain atom denoting falsity, the p_i, q_j, and p are classical ground literals, i.e., either positive atoms or atoms preceded by the classical negation sign $\neg$ [10], that stands for a strong declaration that speaks for itself, and *not* denotes *negation-by-failure*, or in other words, a flop in proving a given statement, once it was not declared explicitly. Under this formalism, every program is associated with a set of *abducibles* [12, 13], given here in the form of exceptions to the extensions of the predicates that make the program, i.e., clauses of the form:

$$exception_{p_1} \cdots exception_{p_j} (0 \leq j \leq k), \qquad (being\ k\ an\ integer\ number)$$

that stand for information or knowledge that cannot be ruled out. On the other hand, clauses of the type:

$$?(p_1, \cdots, p_n, not\ q_1, \cdots, not\ q_m)(n, m \geq 0)$$

also named invariants or restrictions to complain with the universe of discourse, set the context under which it may be understood. The term $scoring_{value}$ stands for the relative weight of the extension of a specific *predicate* with respect to the extensions of peers ones that make the inclusive or global program.

In order to evaluate the knowledge that may be associated to a logic program, an assessment of the *Quality-of-Information (QoI)*, given by a truth-value in the interval $0, \ldots, 1$, that branches from the extensions of the predicates that make a program, inclusive in dynamic environments, is set [15, 16]. On the other hand, a measure of one's confidence that the argument values or attributes of the terms that make the extension of a given predicate, with relation to their domains, fit into a given interval, is also considered, and labeled as *Degree of Confidence (DoC)* [17]. The *DoC* is evaluated as described in [17] and computed using $DoC = \sqrt{1 - \Delta l^2}$, where Δl stands for the argument interval length, which was set to the interval $0, \ldots, 1$. Thus, the universe of discourse is engendered according to the information presented in the extensions of such predicates, according to productions of the type:

$$predicate_i - \bigcup_{1 \leq i \leq m} clause_j((QoI_{x_1}, DoC_{x_1}), \cdots, (QoI_{x_m}, DoC_{x_m})) :: QoI_i :: DoC_i \quad (1)$$

where $\cup$ and m stand, respectively, for *set union* and the *cardinality* of the extension of *predicate$_i$*. *QoI$_i$* and *DoC$_i$* stand for themselves.

3   Case Study

In order to develop a predictive model to assess the pesticides toxicological risk a knowledge database was set, and built around the pesticides records of the *National Pesticide Information Center* [18]. For each pesticide it was considered information regarding environmental fate, human exposure and toxicity (i.e., acute and chronic) both in qualitative and quantitative terms. This section demonstrates how the information comes together and how it is processed.

3.1   Qualitative Data Pre-processing

Aiming at the quantification of the qualitative information and in order to make easy the understanding of the process, it was decided to put it in a graphical form. Taking as an example a set of 3 (three) issues regarding a particular subject (where the possible alternatives are *none*, *low*, *moderate*, *high* and *very high*), a unitary radius circle split into 3 (three) slices is itemized (Fig. 1). The marks in the axis correspond to each of the possible choices. If the answer to issue 1 is *high* the area correspondent is $\pi \times 0.75^2/3$, i.e., 0.19π (Fig. 1(a)). Assuming that in the issue 2 are chosen the alternatives *high* and *very high*, the correspondent area ranges in the interval $\pi \times 0.75^2/3 \cdots \pi \times 1^2/3$, i.e., $0.19\pi \cdots 0.33\pi$ (Fig. 1(b)). Finally, in issue 3 if no alternative is ticked, all the hypotheses should be considered and the area varies in the interval $0 \cdots \pi \times 1^2/3$, i.e., $0 \cdots 0.33\pi$ (Fig. 1(c)). The total area is the sum of the partial ones and is set in the interval $0.38\pi \cdots 0.85\pi$ (Fig. 1(d)). The normalized area is the ratio between the area of the figure and the area of the unitary radius circle. Thus, the quantitative value regarding the subject in analysis is set to the interval $0.38 \cdots 0.85$.

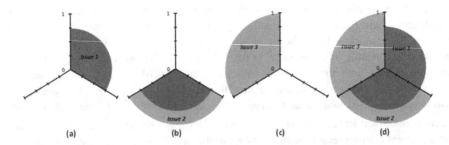

Fig. 1. A view of the qualitative evaluation process.

3.2   A Logic Programming Approach to Data Processing

It is now possible to build up a knowledge database given in terms of the extensions of the relations (or tables) depicted in Fig. 2, which denote a situation where one has to manage information in order to evaluate the *Pesticide Toxicological Risk*. Under this scenario some incomplete and/or default data is present. For instance, in the former case the *ADI* is unknown (depicted by the symbol ⊥), while the *Acute Toxicity* for *Mice/Rats* is not conclusive (*Slightly/Moderate*).

The *Human Exposure* table is populated with 0 (zero) that stands for absence, 1 (one) that denotes food or drinking water only (in *dietary* column), and dermal or inhalation exposure only (in *occupational* column), and 2 (two) stand for simultaneous exposition. The issues presented in *Environmental Fate* table are populated with *absence, low, medium, high* and *very high*, while the columns of *Acute* and *Chronic Toxicity* tables are filled with *absence, slightly, medium, high* and *very high*. In order to quantify the information present in these tables the procedures already described above were followed.

Applying the algorithm presented in [17] to the table or relation's fields that make the knowledge base for *Pesticide Toxicological Risk Assessment* (Fig. 2), and looking to the *DoCs* values obtained as described in [17], it is possible to set the arguments of the predicate *toxicological risk assessment* (*tra*) referred to below, whose extensions denote the objective function with respect to the problem under analyze:

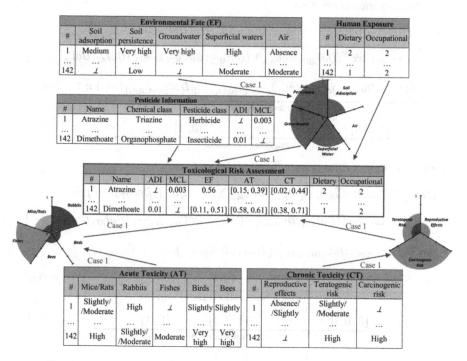

Fig. 2. A fragment of the knowledge base for *Toxicological Risk Assessment*.

$$tra : A_{cceptable}D_{aily}I_{ntake}, M_{aximum}C_{ontamination}L_{evel}, E_{nvironmental}F_{ate}, A_{cute}$$
$$T_{oxicity}, C_{hronic}T_{oxicity}, D_{ietary}H_{uman}E_{xposure}, O_{ccupational}E_{xposure} \rightarrow \{0, 1\}$$

where 0 (zero) and 1 (one) denote, respectively, the truth values *false* and *true*.

The algorithm presented in [17] encompasses different phases. In the former one the clauses or terms that make extension of the predicate under study are established. In the subsequent stage the arguments of each clause are set as continuous intervals. In a third step the boundaries of the attributes intervals are set in the interval [0, 1] according to a normalization process given by the expression $(Y - Y_{min})/(Y_{max} - Y_{min})$, where the Y_s stand for themselves. Finally, the *DoC* is evaluated as described in Sect. 2.

Exemplifying the application of the algorithm presented in [17], in relation to the term (clause) that presents the feature vector $ADI = 0.01$, $MCL = \perp$, $EF = 0.28$, $AT = [0.64, 0.81]$, $CT = [0.04, 0.06]$, $DHE = 1$, $OE = 2$, one may have:

Begin

<<DoCs evaluation>>

<<The predicate's extension that sets the Universe-of-Discourse for the term under observation is fixed>>

$\{\neg\ tra\ \big((QoI_{ADI}, DoC_{ADI}), (QoI_{MCL}, DoC_{MCL}), \cdots, (QoI_{AT}, DoC_{AT}), \cdots\big)$

$\qquad \leftarrow not\ tra\ \big((QoI_{ADI}, DoC_{ADI}), (QoI_{MCL}, DoC_{MCL}), \cdots, (QoI_{AT}, DoC_{AT}), \cdots\big)$

$tra\ \underbrace{\big((1_{0.01}, DoC_{0.01}), (1_{\perp}, DoC_{\perp}), \cdots, (1_{[0.64,\, 0.81]}, DoC_{[0.64,\, 0.81]}), \cdots\big)}_{attribute's\ values} :: 1 :: DoC$

$\qquad \underbrace{[0,\ 0.5] \qquad\quad [0,\ 0.1] \quad\cdots \qquad\qquad [0,\ 1] \qquad\qquad \cdots}_{attribute's\ domains} \qquad\quad \}:: 1$

<<The attribute's values ranges are rewritten>>

$\{\neg\ tra\ \big((QoI_{ADI}, DoC_{ADI}), (QoI_{MCL}, DoC_{MCL}), \cdots, (QoI_{AT}, DoC_{AT}), \cdots\big)$

$\qquad \leftarrow not\ tra\ \big((QoI_{ADI}, DoC_{ADI}), (QoI_{MCL}, DoC_{MCL}), \cdots, (QoI_{AT}, DoC_{AT}), \cdots\big)$

$tra\ \big((1_{[0.01,\, 0.01]}, DoC_{[0.01,\, 0.01]}), (1_{[0,\, 0.1]}, DoC_{[0,\, 0.1]}), \cdots,$

$\qquad\qquad\qquad\qquad\qquad \underbrace{(1_{[0.64,\, 0.81]}, DoC_{[0.64,\, 0.81]}), \cdots\big)}_{attribute's\ values\ ranges} :: 1 :: DoC$

$\qquad \underbrace{[0,\ 0.5] \qquad\qquad\quad [0,\ 0.1] \quad\cdots\quad [0,\ 1] \qquad\qquad \cdots}_{attribute's\ domains} \qquad\quad \}:: 1$

<<*The attribute's boundaries are set to the interval [0, 1]*>>

$\{ \; \neg tra \, ((QoI_{ADI}, DoC_{ADI}), (QoI_{MCL}, DoC_{MCL}), \cdots, (QoI_{AT}, DoC_{AT}), \cdots)$

$\quad \leftarrow not \; tra \, ((QoI_{ADI}, DoC_{ADI}), (QoI_{MCL}, DoC_{MCL}), \cdots, (QoI_{AT}, DoC_{AT}), \cdots)$

$tra \, \Big(\big(1_{[0.02,\,0.02]}, DoC_{[0.02,\,0.02]} \big), \big(1_{[0,\,1]}, DoC_{[0,\,1]} \big), \cdots ,$

$$\underbrace{\big(1_{[0.64,\,0.81]}, DoC_{[0.64,\,0.81]} \big), \cdots \Big)}_{\text{attribute's values ranges once normalized}} :: 1 :: DoC$$

$$\underbrace{[0, 1] \qquad\qquad [0, 1] \quad \cdots \quad [0, 1] \qquad \cdots}_{\text{attribute's domains once normalize}} \quad \}:: 1$$

<<*The DoC's values are evaluated*>>

$\{ \; \neg tra \, ((QoI_{ADI}, DoC_{ADI}), (QoI_{MCL}, DoC_{MCL}), \cdots, (QoI_{AT}, DoC_{AT}), \cdots)$

$\quad \leftarrow not \; tra \, ((QoI_{ADI}, DoC_{ADI}), (QoI_{MCL}, DoC_{MCL}), \cdots, (QoI_{AT}, DoC_{AT}), \cdots)$

$$tra \underbrace{\big((1, 1), \quad (1, 0), \cdots, \quad (1, 0.98), \cdots \big)}_{\substack{\text{attribute's quality−of−information} \\ \text{and respective confidence values}}} :: 1 :: 0.86$$

$$\underbrace{[0.02, 0.02] \quad [0, 1] \quad \cdots \quad [0.64, 0.81] \cdots}_{\text{attribute's values ranges once normalized}}$$

$$\underbrace{[0, 1] \qquad [0, 1] \quad \cdots \quad [0, 1] \quad \cdots}_{\text{attribute's domains once normalized}} \qquad \}:: 1$$

End

4 Artificial Neural Networks

On the one hand, *ANNs* denote a set of connectionist models inspired in the behaviour of the human brain. In particular, the *MultiLayer Perceptron (MLP)* model stands for the most popular *ANN* architecture, where neurons are grouped in layers and only forward connections are set [19]. This provides a powerful base-learner with some advantages with respect to other approaches (e.g., adaptability, robustness, flexibility, nonlinear mapping and noise tolerance), a reason why they are increasingly used in data mining, namely due to its good behaviour in terms of predictive knowledge [20]. The interest in *MLPs* was stimulated by the advent of the *Backpropagation* algorithm in 1986, and since then several fast gradient based variants have been proposed (e.g., *RPROP*) [21]. Yet, these training algorithms minimize an error function by tuning the modifiable parameters of a fixed architecture, which needs to be set a priori. The *MLP* performance will be sensitive to this choice, i.e., a small network will provide limited learning capabilities, while a large one will induce generalization loss (i.e., over fitting). *MLP* is molded on three or more layers of artificial neurons, including an input layer, an output layer and a number of hidden layers with a certain number of active neurons. In addition, there is also a bias, which is only connected to neurons in the hidden and output layers [19]. The correct design of the *MLP* topology is a complex and crucial task, commonly addressed by trial-and-error procedures (e.g., exploring different number of hidden nodes), in a blind search strategy, which only goes through a small set of possible configurations. More elaborated methods have also been proposed, such as pruning [22] and constructive [23] algorithms, although these perform hill-climbing and are thus prone to local minima [11]. The number of nodes in the input layer sets the number of independent variables, and the number of nodes in the output layer denotes the number of dependent ones [19].

On the other hand, the framework presented previously shows how the information comes together and how it is processed. In this section, a data mining approach to deal with the processed information is considered. A hybrid computing approach was set to model the universe of discourse, where the computational part is based on *ANNs*, whose behavior was referred to above, and used not only to structure data but also to capture the problem(s) objective function's nature (i.e., the relationships between inputs and outputs) [24, 25].

Figure 3 shows a case being submitted to the *Pesticide Toxicological Risk Assessment* model. The normalized values of the interval boundaries and its *QoI's* and *DoC's* stand for the inputs to the *ANN*. The output is given in terms of *Pesticide Toxicological Risk* evaluation and the degree of confidence that one has on such a happening. In this study 142 pesticides were considered (i.e., one hundred and forty two terms or clauses of the extension of predicate *tra*). To implement the evaluation mechanisms and to test the model, ten folds cross validation were applied [19]. The back propagation algorithm was used in the learning process of the *MLP*. As the output function in the pre-processing layer it was used the identity one, while in the other layers we considered the sigmoid.

A common tool to evaluate the results presented by the classification models is the coincidence matrix, a matrix of size $L \times L$, where L denotes the number of possible classes. This matrix is created by matching the predicted and target values. L was set to 2 (two) in the present case. Table 1 presents the coincidence matrix of the *ANN* model, where the values presented denote the average of 25 (twenty five) experiments. A glance at Table 1 shows that the model accuracy was 93.7 % (133 instances correctly classified in 142). Therefore, the predictions made by the *ANN* model are satisfactory, attaining accuracies higher than 90 %.

Based on coincidence matrix it is possible to compute sensitivity, specificity, *Positive Predictive Value* (*PPV*) and *Negative Predictive Value* (*NPV*) of the classifier. Briefly, sensitivity evaluates the proportion of true positives that are correctly identified as such, while specificity translates the proportion of true negatives that are correctly identified. *PPV* stands for the proportion of cases with positive results which are correctly classified while *NPV* is the proportion of cases with negative results which are successfully labeled. The values obtained for sensitivity, specificity, *PPV* and *NPV* were 94.7 %, 91.5 %, 95.7 % and 89.6 %, respectively. On the one hand, the proposed model correctly identified 94.7 % of the positive cases, i.e., pesticides with *potential*

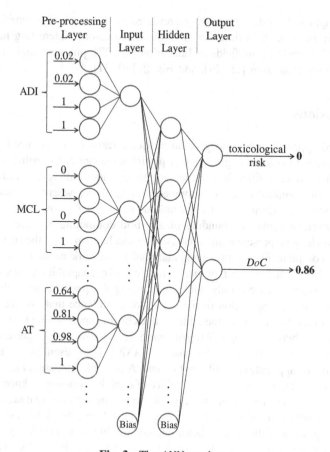

Fig. 3. The *ANN* topology

Table 1. The coincidence matrix for *ANN* model.

Target	Predictive	
	True (1)	False (0)
True (1)	91	4
False (0)	5	42

toxicological risk. On the other hand, it also classified appropriately 91.5 % of the negative cases, i.e., pesticides with *low toxicological risk*.

The present model, beyond to consider the pesticide chemical properties, enables the integration of *Acute* and *Chronic Toxicity* data with other factors such as *Environmental Fate* and *Human Exposure*, being therefore assertive in the prediction of *Pesticide Toxicological Risk*. Thus, it is our claim that the proposed model is able to evaluate the *Toxicological Risk* properly and can be a major contribution to achieve high levels regarding public health protection and environmental sustainability.

The *LP* approach to data processing presented in this work is a generic one, and therefore may be applied in different grounds. Indeed, some interesting results have been obtained, namely in the fields of Education [26, 27], pharmacological properties of Essential Oils evaluation [28, 29], and Health [30, 31].

5 Conclusions

The proposed approach is able to give an adequate response to the need for a good response to predict the toxicological risk of pesticide exposure. Nevertheless, it can be considered a hard task since it is necessary to consider different variables and/or conditions with complex relations entwined among them, where the data may be incomplete, self-contradictory, and even unknown. In order to overcome these difficulties this work presents the founding of a hybrid computing approach that uses a powerful knowledge representation and reasoning mechanism to set the structure of the information, complemented with a computational framework based on *ANNs*, which have been selected due to their proper dynamics, like adaptability, robustness, and flexibility. This approach not only allows evaluating the pesticide toxicological risk, but it also permits the estimation of the degree-of-confidence that one has on such a happening. In fact, this is one of the added values of this approach that arises from the complementarily between *Logic Programming* (for knowledge representation and reasoning) and the computing process based on *ANNs*. The present model is a generic one, susceptive of application in different arenas. A possible limitation on its use is not on the model in itself, but on the unavailability of data, information or knowledge; but, even in these situations, once it has the capacity to handle incomplete data, information or knowledge, either in its qualitative or quantitative form, its usefulness is assured. Future developments of the model should include the biotransformation pathways and routes of exposure, and consider the contact time and the individual's susceptibility. Furthermore, this problem might be approached using others computational frameworks like Case Based Reasoning [27], Genetic Programming [14], or Particle Swarm [32], just to name a few.

Acknowledgments. This work has been supported by COMPETE: POCI-01-0145-FEDER-007043 and FCT – Fundação para a Ciência e Tecnologia within the Project Scope: UID/CEC/00319/2013.

References

1. Hodgson, E.: Introduction to pesticide biotransformation and disposition. In: Hodgson, E. (ed.) Pesticide Biotransformation and Disposition, pp. 1–3. Elsevier, Amsterdam (2012)
2. Needham, L.L., Patterson, D.G., Barr, D.B., Grainger, J., Calafat, A.M.: Uses of speciation techniques in biomonitoring for assessing human exposure to organic environmental chemicals. Anal. Bioanal. Chem. **381**, 397–404 (2005)
3. Environmental Protection Agency: General Principles for Performing Aggregate Exposure and Risk Assessments. Item 6043. https://www.epa.gov/sites/production/files/2015-07/documents/aggregate.pdf

4. Renwick, A.G.: Pesticide residue analysis and its relationship to hazard characterisation (ADI/ARfD) and intake estimations (NEDI/NESTI). Pest Manag. Sci. **58**, 1073–1082 (2002)
5. Esteban, M., Castaño, A.: Non-invasive matrices in human biomonitoring: a review. Environ. Int. **35**, 438–449 (2009)
6. Angerer, J., Ewers, U., Wilhelm, M.: Human biomonitoring: state of the art. Int. J. Hyg. Environ. Health **210**, 201–228 (2007)
7. Antón, A., Castells, F., Montero, J.I., Huijbregts, M.: Comparison of toxicological impacts of integrated and chemical pest management in mediterranean greenhouses. Chemosphere **54**, 1225–1235 (2004)
8. Alister, C., Kogan, M.: ERI: Environmental risk index. A simple proposal to select agrochemicals for agricultural use. Crop Prot. **25**, 202–211 (2006)
9. Juraske, R., Antón, A., Castells, F., Huijbregts, M.A.: PestScreen: a screening approach for scoring and ranking pesticides by their environmental and toxicological concern. Environ. Int. **33**, 886–893 (2007)
10. Neves, J.: A logic interpreter to handle time and negation in logic databases. In: Muller, R., Pottmyer, J. (eds.) Proceedings of the 1984 Annual Conference of the ACM on the 5th Generation Challenge, pp. 50–54. Association for Computing Machinery, New York (1984)
11. Cortez, P., Rocha, M., Neves, J.: Evolving time series forecasting ARMA models. J. Heuristics **10**, 415–429 (2004)
12. Kakas, A., Kowalski, R., Toni, F.: The role of abduction in logic programming. In: Gabbay, D., Hogger, C., Robinson, I. (eds.) Handbook of Logic in Artificial Intelligence and Logic Programming, vol. 5, pp. 235–324. Oxford University Press, Oxford (1998)
13. Pereira, L.M., Anh, H.T.: Evolution prospection. In: Nakamatsu, K., Phillips-Wren, G., Jain, L.C., Howlett, R.J. (eds.) New Advances in Intelligent Decision Technologies. SCI, vol. 199, pp. 51–64. Springer, Berlin (2009)
14. Neves, J., Machado, J., Analide, C., Abelha, A., Brito, L.: The halt condition in genetic programming. In: Neves, J., Santos, M.F., Machado, J.M. (eds.) EPIA 2007. LNCS (LNAI), vol. 4874, pp. 160–169. Springer, Heidelberg (2007)
15. Lucas, P.: Quality checking of medical guidelines through logical abduction. In: Coenen, F., Preece, A., Mackintosh, A. (eds.) Research and Development in Intelligent Systems XX, pp. 309–321. Springer, London (2003)
16. Machado, J., Abelha, A., Novais, P., Neves, J., Neves, J.: Quality of service in healthcare units. In: Bertelle, C., Ayesh, A. (eds.) Proceedings of the ESM 2008, pp. 291–298. Eurosis – ETI Publication, Ghent (2008)
17. Fernandes, F., Vicente, H., Abelha, A., Machado, J., Novais, P., Neves, J.: Artificial neural networks in diabetes control. In: Proceedings of the 2015 Science and Information Conference (SAI 2015), pp. 362–370. IEEE Edition (2015)
18. National Pesticide Information Center. http://npic.orst.edu/index.html
19. Haykin, S.: Neural Networks and Learning Machines. Pearson Education, New Jersey (2009)
20. Mitra, S., Pal, S., Mitra, P.: Data mining in soft computing framework: a survey. IEEE Trans. Neural Netw. **13**, 3–14 (2002)
21. Riedmiller, M.: Advanced supervised learning in multilayer perceptrons—from backpropagation to adaptive learning algorithms. Comput. Stand. Interfaces **16**, 265–278 (1994)
22. Thimm, G., Fiesler, E.: Evaluating pruning methods. In: Proceedings of the International Symposium on Artificial Neural Networks, pp. 20–25. National Chiao-Tung University Edition (1995)
23. Kwok, T., Yeung, D.: Constructive algorithms for structure learning in feedforward neural networks for regression problems: a survey. IEEE Trans. Neural Netw. **8**, 630–645 (1997)

24. Vicente, H., Couto, C., Machado, J., Abelha, A., Neves, J.: Prediction of water quality parameters in a reservoir using artificial neural networks. Int. J. Des. Nat. Ecodyn. **7**, 309–318 (2012)

25. Vicente, H., Dias, S., Fernandes, A., Abelha, A., Machado, J., Neves, J.: Prediction of the quality of public water supply using artificial neural networks. J. Water Supply: Res. Technol. – AQUA **61**, 446–459 (2012)

26. Figueiredo, M., Neves, J., Vicente, H.: A soft computing approach to quality evaluation of general chemistry learning in higher education. In: Caporuscio, M., De la Prieta, F., Di Mascio, T., Gennari, R., Rodríguez, J.G., Vittorini, P. (eds.) Methodologies and Intelligent Systems for Technology Enhanced Learning. Advances in Intelligent and Soft Computing, vol. 478, pp. 81–89. Springer International Publishing, Cham (2016)

27. Neves, J., Figueiredo, M., Vicente, L., Vicente, H.: A case based reasoning view of school dropout screening. In: Kim, K.J., Joukov, N. (eds.) Information Science and Applications. LNEE, vol. 376, pp. 953–964. Springer, Singapore (2016)

28. Neves, J., Martins, M.R., Candeias, F., Arantes, S., Piteira, A., Vicente, H.: An assessment of pharmacological properties of *schinus* essential oils – a soft computing approach. In: Proceedings 30th European Conference on Modelling and Simulation (ECMS 2016), pp. 107–113. European Council for Modelling and Simulation Edition (2016)

29. Neves, J., Martins, M.R., Candeias, F., Ferreira, D., Arantes, S., Cruz-Morais, J., Gomes, G., Macedo, J., Abelha, A., Vicente, H.: Logic programming and artificial neural networks in pharmacological screening of schinus essential oils. Int. J. Biol. Biomol. Agric. Food Biotechnol. Eng. **9**, 706–711 (2015). World Academy of Science, Engineering and Technology, International Science Index 103

30. Vilhena, J., Vicente, H., Martins, M.R., Grañeda, J., Caldeira, F., Gusmão, R., Neves, J., Neves, J.: Antiphospholipid syndrome risk evaluation. In: Rocha, Á., Correia, A.M., Adeli, H., Reis, L.P., Teixeira, M.M. (eds.) New Advances in Information Systems and Technologies. Advances in Intelligent Systems and Computing, vol. 444, pp. 157–167. Springer International Publishing, Cham (2016)

31. Neves, J., Martins, M.R., Vilhena, J., Neves, J., Gomes, S., Abelha, A., Machado, J., Vicente, H.: A soft computing approach to kidney diseases evaluation. J. Med. Syst. **39**, 131 (2015). doi:10.1007/s10916-015-0313-4

32. Mendes, R., Kennedy, J., Neves, J.: Watch thy neighbor or how the swarm can learn from its environment. In: Proceedings of the 2003 IEEE Swarm Intelligence Symposium (SIS 2003), pp. 88–94. IEEE Edition (2003)

Classification of Brain Tumor MRIs
Using a Kernel Support Vector Machine

Mahmoud Khaled Abd-Ellah[1], Ali Ismail Awad[2,3](✉), Ashraf A.M. Khalaf[4],
and Hesham F.A. Hamed[4]

[1] Electronic and Communication Department,
Al-Madina Higher Institute for Engineering and Technology, Giza, Egypt
eng_mahmoudkhaled@yahoo.com
[2] Department of Computer Science, Electrical and Space Engineering,
Luleå University of Technology, Luleå, Sweden
ali.awad@ltu.se
[3] Faculty of Engineering, Al Azhar University, Qena, Egypt
[4] Electrical Engineering Department, Faculty of Engineering,
Minia University, Minia, Egypt
ashkhalaf@yahoo.com, hfah66@yahoo.com

Abstract. The use of medical images has been continuously increasing, which makes manual investigations of every image a difficult task. This study focuses on classifying brain magnetic resonance images (MRIs) as normal, where a brain tumor is absent, or as abnormal, where a brain tumor is present. A hybrid intelligent system for automatic brain tumor detection and MRI classification is proposed. This system assists radiologists in interpreting the MRIs, improves the brain tumor diagnostic accuracy, and directs the focus toward the abnormal images only. The proposed computer-aided diagnosis (CAD) system consists of five steps: MRI preprocessing to remove the background noise, image segmentation by combining Otsu binarization and K-means clustering, feature extraction using the discrete wavelet transform (DWT) approach, and dimensionality reduction of the features by applying the principal component analysis (PCA) method. The major features were submitted to a kernel support vector machine (KSVM) for performing the MRI classification. The performance evaluation of the proposed system measured a maximum classification accuracy of 100 % using an available MRIs database. The processing time for all processes was recorded as 1.23 seconds. The obtained results have demonstrated the superiority of the proposed system.

Keywords: Brain tumor · MRIs classification · K-means · DWT · PCA · KSVM

1 Introduction

A brain tumor is an abnormal growth of tissue inside or around the brain that can disrupt proper brain function and that creates increasing pressure. According to the National Brain Tumor Foundation (NBTF), the number of people in

© Springer International Publishing Switzerland 2016
H. Li et al. (Eds.): WIS 2016, CCIS 636, pp. 151–160, 2016.
DOI: 10.1007/978-3-319-44672-1_13

developed countries who die as a result of brain tumors has increased by 300 % over the past three decades [1,2]. Magnetic resonance imaging is an advanced medical imaging technique that provides rich information about the anatomy of human soft tissues. It has several advantages over other imaging techniques. The wide spread of brain tumors has led to the production of a considerable amount of medical images. However, some physicians still manually investigate MRIs, which is less accurate and a time-consuming task [3,4].

Efficiently and accurately investigating MRIs requires a computerized system in which all the processes should be automatically performed. Although a medical image classification system was proposed in [5], it takes a general perspective and does not specifically focus on brain tumor MRIs. The development of a complete computer-aided diagnosis (CAD) system for brain tumor detection is not an easy task. It involves several processes, such as image enhancement, feature extraction and selection, feature reduction, and feature classification. Considering the presence and absence of brain tumors in MRI classification adds new challenges to the system.

In this study, a conceptual design and implementation of a complete (CAD) system for brain tumor detection and MRI classification is presented. The objective of the presented system is to detect brain tumors and classify MRIs into normal (brain tumor absent) and abnormal (brain tumor present), where the normal image can be excluded as early as possible. The system is considered to be an early prerequisite phase for brain tumor detection, segmentation, and tracking such that all focus can be directed to abnormal MRIs.

Although several methods for brain tumor detection and segmentation can be found in the literature [6–13], the contribution of this study is the development of a complete framework for detecting the presence and absence of brain tumors in MRIs, whereas most of the state-of-the-art methods assume only the presence of brain tumors. The proposed system applies kernel support vector machine (KSVM) for tumor detection and MRI classification, which can be treated as a machine learning and pattern recognition problem. However, SVM is not a new method in brain tumor detection [14–16], but including it in one CAD system for different application contexts enriches the presented contributions. The proposed system assists radiologists in interpreting MRIs, improves the brain tumor diagnostic accuracy, and directs the processing toward the abnormal images.

The remainder of this paper is organized as follows. The description of the conceptual design and the implementation of the proposed system are provided in Sect. 2. Section 3 offers detailed information on the performance evaluation, the database, and the results obtained using the proposed CAD system. Finally, concluding remarks and future work are outlined in Sect. 4.

2   Design and Implementation

Computer-aided diagnosis (CAD) plays an important role in medical diagnosis, where its accuracy and reliability are crucial. The design of the detection and classification system proposed in this study covers all the necessary phases,

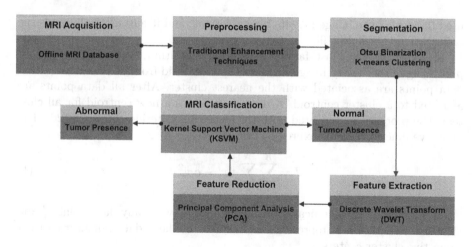

Fig. 1. Block diagram of the proposed CAD classification system. All processes are included in the system, starting with MRI acquisition and ending with MRI classification.

from image acquisition to MRI classification, to group MRIs into normal and abnormal images. A block diagram that shows all phases of the proposed system is presented in Fig. 1. In the proposed CAD system, the MRI is provided to the system as an off-line input, and then preprocessing and region of interest (ROI) segmentation are applied to the image. Subsequently, feature extraction is applied to the segmented image, and feature reduction is then employed for selecting the relevant features for the final classification step.

Although brain MRI is efficient and widely used because it provides more details that aid in brain tumor diagnosis, it is difficult to obtain brain MRIs due to privacy issues. Thus, off-line MRIs were used in the image acquisition stage of the system. The image database was obtained from the Harvard Medical School, MICCAI 2014 Machine Learning Challenge (MLC), and Laboratory for Computational Imaging Biomarkers websites. Detailed information on the used database is presented in Sect. 3.1.

Magnetic resonance images can be affected by several types of noise and suffer from resolution degradation. A preprocessing stage was considered as a part of the CAD system to improve the image quality by reducing the noise level and enhancing the overall image resolution [17]. Several conventional filters could be applied in this stage; however, a median filter was selected due its efficiency in image de-noising and filtering [18].

2.1 Segmentation of Region of Interest

Region of interest (ROI) segmentation is an important step in the process chain that divides the MRI into regions. The segmentation stage was performed in two steps: Otsu binarization and K-means clustering. Otsu binarization was used to convert the image into its binary format, and it automatically finds the binarization threshold [19].

K-means, as the most famous clustering algorithm, was used for further processing of the binary image. It takes the k centroid from a set of data. Other data points are associated with the nearest cluster. After all data points are allocated to a cluster centroid, K-means calculates a new centroid for all clusters. The process was repeated until the clustering criterion converges [20]. The objective function used is expressed in Eq. 1.

$$J = \sum_{j=1}^{K} \sum_{i=1}^{N} \|x_i^j - c_j\|^2 \tag{1}$$

where $\|x_i^j - c_j\|^2$ is an indicator of the distance between any data point x_i^j and the cluster center c_j. It represents the distance of the n data points from their respective cluster centers.

2.2   Feature Extraction

In this stage, the discrete wavelet transform (DWT) technique was applied to the segmented image to extract the features. DWT works by first converting images from the spatial domain to then frequency domain. Then, the actual DWT is performed by filtering the image using two filters, a low pass filter and a high pass filter, in both the vertical and horizontal directions. In every DWT level, the image is divided into four coefficients: LL, LH, HL, and HH. The LL sub-bands come from using the low pass filter in the horizontal direction and the low pass filter in the vertical direction, and these sub-bands are known as the approximation coefficient. The other sub-bands are known as detailed coefficients. In the proposed design, a level-3 decomposition via the Daubechies (DB) wavelet family was used to extract features from the approximation coefficient [21].

2.3   Feature Reduction

Extra unnecessary features will increase the classification complexity, prolong the computational time, and require more storage memory. Thus, feature reduction was considered as a part of the CAD design. The principal component analysis (PCA) method was used to reduce the dimension of data according to their importance and variance. The PCA method makes the components of the input feature set perpendicular, and then it rearranges them in terms of the highest variation. The components with a low variation in the feature set are removed. A normalization process on the input feature set was performed to obtain a zero mean and identity variance prior to applying the PCA method [22].

2.4   Training and Classification

A kernel support vector machine (KSVM) was finally applied for MRI classification. Although there are several families of SVM, such as autoscale, box constraint, and kernel, the kernel support vector machine was selected due to its

Table 1. Distribution of images in the evaluation database.

Total images	No. of images in the training set		No. of images in the testing set	
	Normal	Abnormal	Normal	Abnormal
80	5	43	5	27

wide range of kernel functions, such as linear, polynomial, and Gaussian radial basis function (GRB). The following kernel functions were considered in the CAD system design:

Linear kernel: It is the simplest kernel and is defined as:

$$K(x_i, x_j) = x_i^T x_j \tag{2}$$

Polynomial kernel: It is suited for problems with normalized training data and is defined as:

$$K(x_i, x_j) = (\alpha x_i^T x_j + c)^d \tag{3}$$

Gaussian radial basis function (GRB) kernel, which is defined as:

$$K(x_i, x_j) = \exp(-\frac{\|x_i - x_j\|^2}{2\sigma^2}) \tag{4}$$

where σ is an adjustable parameter. It is a non-linear kernel and is very sensitive to noise. Multilayer perceptron (MLP) kernel, also called a hyperbolic tangent kernel or sigmoid kernel, and it is defined as:

$$K(x_i, x_j) = \tanh(\gamma x_i^T x_j + r) \tag{5}$$

where γ is the slope and r is the intercept constant.

In KSVM, the kernel parameters need to be adjusted prior to the data training process. During our experiments, we applied different types of kernel functions and determined that the Gaussian radial basis function (GRB) kernel with a default scaling factor is the best [14]. The training dataset contains normal and abnormal sub-sets, which allow the classification step to recognize the type of the new MRI as a normal or abnormal image.

3 Performance Evaluation

3.1 Testing Environment

To evaluate the proposed algorithm based on the confusion matrix, we used the following metrics: **sensitivity**, which represents the proportion of actual positives that are correctly identified; **specificity**, which indicates the proportion of negatives that are correctly identified; and **accuracy**, which is the proportion of both true positives and true negatives [23], as given in Eqs. 6, 7, and 8, respectively.

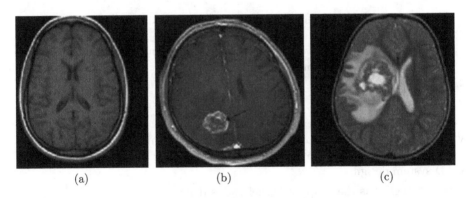

(a)   (b)   (c)

Fig. 2. Sample of the MRI database used in the experimental work. (a) normal image, (b) benign image (abnormal), and (c) malignant image (abnormal).

$$\textbf{Sensitivity} \text{ (True positive rate) } = \frac{TP}{TP + FN} \tag{6}$$

$$\textbf{Specificity} \text{ (True negative rate) } = \frac{TN}{TN + FP} \tag{7}$$

$$\textbf{Accuracy} \text{ (Correctly classified samples (\%)) } = \frac{TN + TP}{TN + TP + FN + FP} \tag{8}$$

where true positives (TPs) are the correctly classified positive cases, true negatives (TNs) are the correctly classified negative cases, false positives (FPs) are the incorrectly classified positive cases, and false negatives (FNs) are the incorrectly classified negative cases.

The evaluation database contains a total of 80 MRIs, including 10 normal and 70 abnormal images. The images were randomly selected because there are 1 type of normal brain and 2 different types of abnormal brain in the image database. The specifications of the utilized database are presented in Table 1. A sample image of the database is shown in Fig. 2. However, the brain tumor type, benign or malignant, was not considered in the classification process. The presence or absence of the tumor was the only basis of the classification.

The proposed technique was developed locally and successfully trained in MATLAB® R2014a using a combination of the wavelet and support vector machine toolboxes. The machine is equipped with an Intel® Core ® i5 processor operating at 2.6 GHz with 6 GB of RAM and Windows® 64-bit operating system.

3.2   Accuracy Analysis

The experimental work began by testing several KSVM kernels for identifying the best classification accuracy, sensitivity, and specificity. The experimental results for normal and abnormal MRI classification are listed in Table 2.

The results in Table 2 show that the RGB kernel obtains quite excellent results on both training and validation images. For the linear kernel, the overall

Table 2. Classification rates for each KSVM kernel. The GRB kernel performs well in all the classification factors.

KSVM kernel	Factors (image)				Evaluation metrics (%)		
	TP	TN	FP	FN	Sensitivity	Specificity	Accuracy
Linear kernel	57	10	13	0	100	43	83.5
MP kernel	50	8	20	2	96	28.5	72.5
Polynomial kernel	69	9	1	1	98.5	90	97.5
GRB kernel	70	10	0	0	100	100	100

Table 3. Comparison of the obtained classification results against the available classification approaches in the literature.

KSVM method	Evaluation metrics (%)		
	Sensitivity	Specificity	Accuracy
PCA + SVM [24]	89	84	85
ICA + SVM [24]	87	75	79
PCC + SVM [24]	89	77	82
DWT + PCA + K-NN [25]	96	97	98
DWT + PCA + ANN [25]	95.9	96	97
FPCNN + DWT+ PCA + FFNN [2]	100	92.8	99
K-means + DWT + PCA + KSVM	**100**	**100**	**100**

classification accuracy was 83.5 %; for the multilayer perceptron (MLP) kernel, it was 72.5 %; for the polynomial kernel, it was 97.5 %; and for the GRB kernel, the classification accuracy was 100 %.

To ensure credibility of the obtained results, a comparison with results in the literature was performed. The comparison outcome of the proposed method (GRB kernel) with six popular methods reported in the literature is reported in Table 3.

3.3   Processing Time Analysis

The processing time is another important factor for evaluating the proposed CAD system. The time for SVM training was not considered because the parameters of the SVM were kept unchanged after training. The following strategy was to send all the images into the CAD system, record the corresponding computation time for every stage for every single image, and compute the average value depicting the time consumed by different stages. The average processing time for every stage per every single MRI is shown in Table 4.

Table 4. The average processing times of different stages in the proposed CAD system.

Process	Processing time (in second)
Preprocessing	0.099
Segmentation	0.380
Feature extraction	0.197
Feature reduction	0.213
Classification	0.337

4   Conclusions

Automatically detecting brain tumors from magnetic resonance images (MRIs) using a computer-aided diagnosis (CAD) system remains a challenge. In this paper, a CAD system for detecting and classifying MRIs according to the presence or absence of tumors was presented. The developed CAD system employs K-means, discrete wavelet transform (DWT), principal component analysis (PCA), and kernel support vector machine (KSVM) methods for segmentation, feature detection, feature reduction, and MRI classification, respectively. Four kernels were evaluated for identifying the highest classification accuracy. The experiments showed that the GRB kernel achieved a classification accuracy of 100 % compared to 83.5 % for the linear kernel, 72.5 % for the multilayer perceptron (MLP) kernel, and 97.5 % for the polynomial kernel. Future work will focus on analyzing the complexity of the system and on reducing the processing time.

References

1. Logeswari, T., Karnan, M.: An improved implementation of brain tumor detection using segmentation based on hierarchical self organizing map. Int. J. Comput. Theor. Eng. **2**(4), 591 (2010)
2. El-Dahshan, E.S.A., Mohsen, H.M., Revett, K., Salem, A.B.M.: Computer-aided diagnosis of human brain tumor through MRI: a survey and a new algorithm. Expert Syst. Appl. **41**(11), 5526–5545 (2014)
3. Jayadevappa, D., Srinivas Kumar, S., Murty, D.: Medical image segmentation algorithms using deformable models: a review. IETE Tech. Rev. **28**(3), 248–255 (2011)
4. Yazdani, S., Yusof, R., Karimian, A., Pashna, M., Hematian, A.: Image segmentation methods and applications in MRI brain images. IETE Tech. Rev. **32**(6), 413–427 (2015)
5. Abedini, M., Codella, N.C.F., Connell, J.H., Garnavi, R., Merler, M., Pankanti, S., Smith, J.R., Syeda-Mahmood, T.: A generalized framework for medical image classification and recognition. IBM J. Res. Dev. **59**(2/3), 1–18 (2015)
6. Prastawa, M., Bullitt, E., Moon, N., Van Leemput, K., Gerig, G.: Automatic brain tumor segmentation by subject specific modification of atlas priors. Acad. Radiol. **10**(12), 1341–1348 (2003)
7. Prastawa, M., Bullitt, E., Ho, S., Gerig, G.: A brain tumor segmentation framework based on outlier detection. Med. Image Anal. **8**(3), 275–283 (2004)

8. Saha, B.N., Ray, N., Greiner, R., Murtha, A., Zhang, H.: Quick detection of brain tumors and edemas: a bounding box method using symmetry. Comput. Med. Imaging Graph. **36**(2), 95–107 (2012)

9. Gordillo, N., Montseny, E., Sobrevilla, P.: State of the art survey on MRI brain tumor segmentation. Magn. Reson. Imaging **31**(8), 1426–1438 (2013)

10. Nabizadeh, N., Kubat, M.: Brain tumors detection and segmentation in MR images: Gabor wavelet vs. statistical features. Comput. Electr. Eng. **45**, 286–301 (2015)

11. Zhang, N., Ruan, S., Lebonvallet, S., Liao, Q., Zhu, Y.: Kernel feature selection to fuse multi-spectral MRI images for brain tumor segmentation. Comput. Vis. Image Underst. **115**(2), 256–269 (2011)

12. Aslam, A., Khan, E., Beg, M.S.: Improved edge detection algorithm for brain tumor segmentation. Procedia Comput. Sci. **58**, 430–437 (2015). Second International Symposium on Computer Vision and the Internet (VisionNet 15)

13. Abdel-Maksoud, E., Elmogy, M., Al-Awadi, R.: Brain tumor segmentation based on a hybrid clustering technique. Egypt. Inf. J. **16**(1), 71–81 (2015)

14. Ayachi, R., Ben Amor, N.: Brain tumor segmentation using support vector machines. In: Sossai, C., Chemello, G. (eds.) ECSQARU 2009. LNCS, vol. 5590, pp. 736–747. Springer, Heidelberg (2009)

15. Bauer, S., Nolte, L.-P., Reyes, M.: Fully automatic segmentation of brain tumor images using support vector machine classification in combination with hierarchical conditional random field regularization. In: Fichtinger, G., Martel, A., Peters, T. (eds.) MICCAI 2011, Part III. LNCS, vol. 6893, pp. 354–361. Springer, Heidelberg (2011)

16. Natteshan, N.V.S., Angel Arul Jothi, J.: Automatic classification of brain MRI images using SVM and neural network classifiers. In: El-Alfy, E.-S., Thampi, S.M., Takagi, H., Piramuthu, S., Hanne, T. (eds.) Advances in Intelligent Informatics. AISC, vol. 320, pp. 19–30. Springer, Heidelberg (2015)

17. Toennies, K.D.: Guide to Medical Image Analysis: Methods and Algorithms. Advances in Computer Vision and Pattern Recognition. Springer Science & Business Media, Heidelberg (2012)

18. Youlian Zhu, C.H.: An improved median filtering algorithm for image noise reduction. In: 2012 International Conference on Solid State Devices and Materials Science, pp. 609–616. Elsevier (2012)

19. Somasundaram, K., Genish, T.: Modified Otsu thresholding technique. In: Balasubramaniam, P., Uthayakumar, R. (eds.) ICMMSC 2012. CCIS, vol. 283, pp. 445–448. Springer, Heidelberg (2012)

20. Cheng, J., Xiaoyun Chen, H.: An enhanced k-means algorithm using agglomerative hierarchical clustering strategy. In: International Conference on Automatic Control and Artificial Intelligence (ACAI 2012), 3–5 March, pp. 407–410. IEEE (2012)

21. Abo-Zahhad, M., Gharieb, R.R., Ahmed, S.M., Abd-Ellah, M.K.: Huffman image compression incorporating DPCM and DWT. J. Signal Inf. Process. **6**, 123–135 (2015)

22. Zhang, Y., Wu, L., Wei, G.: A new classifier for polarimetric SAR images. Prog. Electromagnet. Res. **94**, 83–104 (2009)

23. Kolusheva, S., Yossef, R., Kugel, A., Hanin-Avraham, N., Cohen, M., Rubin, E., Porgador, A.: A novel "reactomics" approach for cancer diagnostics. Sensors **12**(5), 5572–5585 (2012)

24. Wang, H., Fei, B.: A modified fuzzy c-means classification method using a multiscale diffusion filtering scheme. Med. Image Anal. **13**(2), 193–202 (2009). Includes Special Section on Functional Imaging and Modelling of the Heart

25. Arimura, H., Tokunaga, C., Yamashita, Y., Kuwazuru, J.: Magnetic resonance image analysis for brain CAD systems with machine learning. In: Suzuki, K. (ed.) Machine Learning in Computer-Aided Diagnosis: Medical Imaging Intelligence and Analysis, pp. 258–296. IGI Gloabal, Hershey (2012)

Supporting the Viability of E-Health Services with Pattern-Based Business Model Design

The Case of an E-Mental Health App for Maternal Depression

Michaela Sprenger[✉]

University of St. Gallen, St. Gallen, Switzerland
michaela.sprenger@unisg.ch

Abstract. Designing viable business models for e-health services is not trivial as people in charge of business model design often lack the respective knowledge and experience. E-health business model design patterns should support inexperienced business model designers as they document existing business model logics of the e-health domain for reuse. This paper aims at understanding how exactly a pattern-based business model design supports the viability of e-health business models by applying the e-health business model design patterns to a specific e-health service – an e-mental health app for maternal depression. A focus group workshop reveals that these design patterns sensitize the participants to the viability aspects of the business model and thereby help to enhance its viability.

Keywords: Business model design · Design patterns · E-health · Maternal depression · Viability

1 Introduction

Thanks to technological advances, patients and health-conscious people are entering the digital world and use e-health services to share their health-related experiences, to schedule appointments with their doctors, or to get medical advice and treatment [1]. However, many of these e-health services are not sustainable and result in economical failure as they lack a viable business model that defines how the service creates, delivers, and captures sufficient value for all stakeholders involved [2, 3]. This lack of business model considerations does not only lead to discontinued e-health services: Some of these services do not even make it to the market at all and end as successful prototypes [4].

To support the business model design for e-health services, so called e-health business model design patterns have been developed that document business model logics of existing e-health services and make them available for reuse [5]. The aim of this paper is to understand how exactly these patterns support the design of viable business models for e-health services. In order to do so, we conduct a focus group workshop for analyzing a pattern-based business model design by means of a concrete case - an e-mental health app for maternal depression. We deem this e-health service as

H. Li et al. (Eds.): WIS 2016, CCIS 636, pp. 161–175, 2016.
DOI: 10.1007/978-3-319-44672-1_14

suitable for our purpose as extant research shows that such an app would be technically feasible and appreciated by potential end-users [6, 7]. While existing pregnancy related apps focus on physical aspects and neglect the diagnosis and treatment of mental conditions [8], our case requires a distinct and viable business model that has yet to be designed.

2   Background

2.1   E-Mental Health App for Maternal Depression

Maternal depression encompasses depression during pregnancy and within the first year after delivery [9]. Analogously to other depressive conditions, affected women feel hopeless and might be overwhelmed with their role of being a mother [10], which also affects the cognitive and emotional development of the child [11]. Even though 38 % of pregnant women [12] and up to 16 % of women after giving birth [13] are affected, maternal depression often remains undetected because in the majority of cases health professionals do not notice this condition during routine clinical practice [14, 15]. One of the main reasons for not detecting maternal depression is that women are often hesitant when it comes to talking about their emotional distress [6]. Moreover, women after delivery are busy with their baby and live according to the baby's napping and feeding schedules [16]. This makes it even harder for them to actually go to a health appointment. One possible solution would be to offer an e-health service in form of an e-mental health app that supports women in detecting maternal depression and that refers them to a health professional if needed. The aforementioned barriers of detecting maternal depression could be reduced by the app: On the one hand, women would be less inhibited to use such an app as they could reveal their emotional distress in an anonymous way; on the other hand, the app would provide a high level of flexibility as the women could use the app wherever they are whenever they have time [6, 16]. By offering an e-mental health app for maternal depression, the condition could be detected at an early stage and long-term consequences for the women and their babies as well as treatment costs could be significantly reduced [11, 15]. Since apps regarding other mental health conditions already proofed to be useful for detection and treatment purposes [17–20], it is surprising that there is not any app on the market that supports the diagnosis and treatment of maternal depression [6]. A review performed by Osma et al. [8] discovered many pregnancy related apps, however, they were rather focused on physiological aspects than on emotional ones. With the lack of e-mental health apps for maternal depression, a viable business model for this e-health service has yet to be found. This paper develops such a business model of an e-mental health app regarding maternal depression by applying a pattern-based business model design. In order to do so, business model related questions like "Which stakeholders should be involved?", "How is value created and delivered?", and "How will this app be financed?" will be answered.

2.2   Viable Business Model Design for E-Health Services

A business model illustrates how value is created, delivered, and captured by a business [2]. To be viable, the business model has to be designed in such a way that the business creates and delivers value to the relevant customers while at the same time appropriates sufficient value for itself [21]. Designing such a viable business model is not trivial as, in reality, the people in charge of designing a business model for their product or service are rather focused on other topics in their daily life and therefore lack the knowledge and experience regarding business model design [2]. For e-health services, this inexperience often leads to a failure of the business as relevant business model aspects are not considered [22].

To overcome the inexperience, so called business model design patterns can be applied. These design patterns have their origins in the area of architecture where they match recurring problems with suitable architectural design solutions [23]. Even designers with a lack of the respective knowledge can (re-)use the documented solutions from these design patterns whenever they have to solve similar problems. Analogously to design patterns in architecture, business model design patterns document common business model logics that might be reused by other business model designers in similar contexts [2, 24–26]. Since 90 % of business model innovations are recombinations of extant business model patterns, the reuse of existing business model logics seems to be a valid approach to business model design [27].

However, general business model design patterns present business model logics on a rather abstract level [28]. As a consequence, the business model designer would have to transfer this high-level solution to his own specific industry and case. To facilitate this transfer, so called e-health business model design patterns have been introduced that are specific to the e-health domain and focus on the problems, goals, and stakeholders relevant for e-health [5]. Overall, 37 e-health business model design patterns have been developed (cf. Table 1) that are presented in a consistent structure: Each pattern states a problem relevant to the e-health domain and proposes goals that might be pursued while solving this problem, then the pattern is presented as a solution. In order to facilitate understanding, the involved stakeholder groups are highlighted. These stakeholders are subdivided according to Mantzana [29] into providers (e.g. doctors), supporters (e.g. IT providers), receivers (e.g. patients), and controllers (e.g. health authorities). Moreover, an example from the e-health domain that already instantiated the pattern is presented. An illustration of the example clarifies the value flows between the involved actors. Figure 1 presents the exemplary e-health business model design pattern "patient network". This pattern refers to the problem that it is rather difficult for patients to find people with a similar health condition. The proposed solution is a community that connects patients sharing the same condition so that they can exchange their experiences. Thereby, patients get fast and efficient access to one another to enhance their level of information on the specific condition. The platform PatientsLikeMe [30] already implemented this pattern.

The idea of these e-health specific patterns is that unexperienced business model designers can relate to the presented problems, stakeholders, and examples and thus can rather easily transfer the presented solution to their own e-health case. As the example

Table 1. E-health business model design patterns [5]

01 24/7 Telehealth	**14** Franchising	**27** Partnership for trust
02 Access to healthcare abroad	**15** Freemium	**28** Patient engagement system
03 Automation	**16** Full healthcare service provider	**29** Patient network
04 Collective intelligence	**17** Gamification	**30** Pay-per-use
05 Commission-based revenue	**18** Health wearables	**31** Razor and blade
06 Crowdsourcing	**19** Healthcare bartering	**32** Reverse auction
07 Data-based customization	**20** Healthcare crowdfunding	**33** Secure platform
08 Data-based pricing	**21** Healthcare data selling	**34** Subscription-based revenue
09 Data for trust	**22** Digital connectivity	**35** Targeting new segments
10 Direct-to-consumer tests	**23** Lock-in	**36** Third-party channels
11 Expert platform	**24** Marketplace for clinical data	**37** Verified cost transparency
12 Fee for health	**25** Open healthcare ecosystem	
13 Flatrate for health	**26** Partnership for customization	

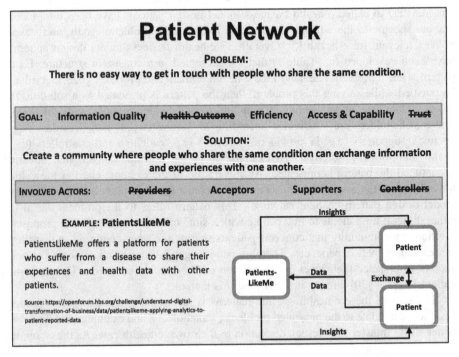

Fig. 1. The e-health business model design pattern "patient network"

shows that the presented solution proofed to be viable in a similar context, each pattern is supposed to support viable business model design.

3 Method

This paper aims at understanding how exactly e-health business model design patterns support the design of viable business models for e-health services by analyzing the pattern-based business model design for an e-mental health app regarding maternal depression. For this purpose a confirmatory focus group was judged to be a suitable method [31]. By conducting a focus group we could get rich data regarding the deployment of the e-health business model design patterns and the consequences for viable business model design [32]. We followed Tremblay et al.'s [31] suggestion to recruit focus group participants that are familiar with the application environment and would be potential users of the e-health business model design patterns. We therefore performed the focus group with employees from a software company that were in charge of designing an app for maternal depression and were in need of a corresponding viable business model.

The focus group was conducted in form of a workshop that followed the steps of a pattern-based business model design as illustrated in Fig. 2 [33].

First, the participants of the focus group were asked to document their current idea of the business model for the e-mental health app in a formalized way as a baseline. Here, they decided to apply the business model canvas [2] to define their business model baseline, because they were already familiar with it. Afterwards they got the task to define the challenges regarding the viability of their current business model idea. For example, if they had the feeling that the business model would not create enough value for one of the relevant stakeholder groups, they were asked to list that as a viability challenge. As a third step, the focus group participants were provided with the e-health business model design patterns, whereas each pattern was printed on a card and each participant got his own set consisting of all 37 e-health business model design patterns. The whole group got a short introduction to the e-health business model design patterns regarding their purpose and structure. Then, the participants were asked to go through the design patterns to come up with ideas how they might solve the viability challenges of their business model baseline by adapting its current design. Each participant wrote his ideas on sticky notes which he afterwards presented to the whole group. After the

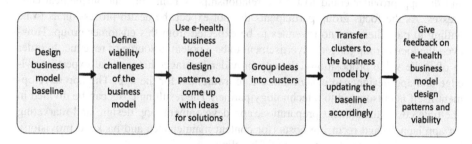

Fig. 2. Structure of a pattern-based business model design

presentation they were asked to group similar ideas into clusters. These clusters were then transferred to the business model by correspondingly updating the baseline. At the end, the focus group participants gave feedback if and how the e-health business model design patterns helped them to address the viability challenges of the baseline and to come up with a more viable business model for their e-mental health app. Here, they were also asked to articulate the assumptions the updated business model is based on and to indicate how realistic these assumptions are [34].

The focus group was conducted by two researchers, whereas one researcher moderated the session and the other one took notes. Here, it was of special interest how the participants applied the patterns, e.g. which patterns they judged to be useful to increase the business model's viability, which ideas they came up with based on those patterns, and how they transferred the idea clusters to their business model.

4 Findings and Discussion

4.1 Business Model Baseline and Its Viability Challenges

The first step of the focus group session was to capture the current idea of the e-mental health app for maternal depression as a business model baseline. Even though it can be described in different ways, the focus group participants opted for the business model canvas [2] to illustrate their business model baseline as they were familiar with this concept. This canvas shows how a business creates value (the value proposition as well as the therefore needed key activities, key resources, and partners), how value is delivered (customers as well as the respective relationships and channels), and how value is captured for the business itself (revenues and costs).

The focus group participants outlined the business model baseline as follows: Regarding the customer group of pregnant women/new mothers the main value proposition lies in offering a way to prevent and detect maternal depression via an easy to use, self-testing tool that protects the women's data. The app provider has an indirect relationship with the women via the health professionals that should use and recommend the app to them. The women can then download the app in a self-service manner via the app store for a one-time fee. With regard to the second customer group, the health professionals, using and recommending the app should enhance the image of an innovation driven health provider as well as support clinical studies focusing on maternal depression. Medical associations should help to create awareness for the app and the app provider could foster the relationship by being present on medical congresses. As the focus group participants did not expect the health professionals to be willing to pay, there are no revenues to be expected from this customer group. However, the focus group judged advertisement to be a potential source of revenue. In order to offer the value propositions, the app provider leverages his medical expert knowledge and app designer to design and create awareness for the app. The app development itself is outsourced to a technology partner. The resulting costs can be divided in one-time costs (for content preparation, app development, app design, and marketing for app launch) and recurring costs (for content maintenance and backend provision). Figure 3 illustrates the business model baseline.

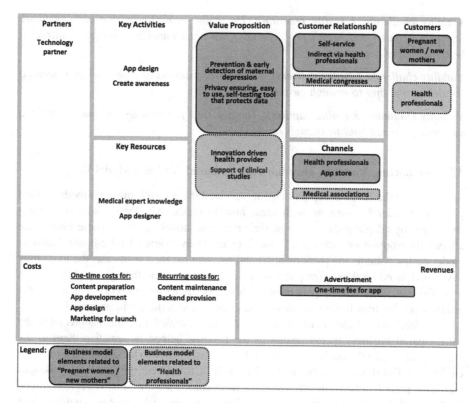

Fig. 3. Business model baseline in form of the business model canvas [2]

After designing the business model baseline, the participants were asked to formulate challenges that might compromise the viability of the e-mental health app. These challenges could refer to the lack of value creation for the different stakeholders (i.e. the value proposition is not compelling); the lack of value delivery (i.e. the value is not provided to (enough) stakeholders), or the lack of value capture (i.e. the app provider cannot appropriate enough of the generated value for himself). Regarding value creation, the focus group doubted that the current value propositions for both the health professionals and the pregnant women/new mothers were convincing enough to attract high numbers within both customer groups. Additionally, the focus group participants were skeptical with regard to the value delivery for the women. Here, the health professionals were designed as a channel as they should use and recommend the app to women during their consultations. However, as health professionals might not be convinced of the value creation, they might not act as a good channel to women, leading to a low adoption rate in the women customer segment. A low adoption from women would lead to a challenge in value capture: With only few women buying the app, the one-time fee as well as the advertising would probably not lead to high enough revenues to cover all costs. Hence, the participants formulated the following viability challenges:

Viability challenge 1 (value creation): The value propositions for both customer segments (women and health professionals) are not compelling enough to attract a high number of customers.

Viability challenge 2 (value delivery): Health professionals as a channel to women do not deliver the app to enough customers.

Viability challenge 3 (value capture): The one-time fee paid by the women and the advertising do not lead to enough revenues.

4.2   Supporting Viability Through Pattern-Based Business Model Design

In a next step the focus group participants received the e-health business model design patterns in order to come up with ideas how to tackle the aforementioned viability challenges by adapting the design of their business model baseline. These ideas were afterwards grouped into clusters. Table 2 gives an overview of all e-health business model design patterns the participants found relevant for their own case, the ideas they developed based on the design patterns, as well as the clusters they grouped their ideas into. This is the core of the pattern-based business model design process as it lays the groundwork for transferring the design patterns to a specific business model.

The feedback of the focus group participants revealed that they appreciated the pattern-based business model design as the patterns helped to sensitize them to the viability aspects of the business model and as a consequence to tackle all three viability challenges. The design patterns fostered ideas how to address the viability challenges which the participants would otherwise not have come up with. This was also enforced by the fact that the design patterns are based on innovative e-health cases and therefore triggered creative, out-of-the box ideas that already proofed to be suitable within one's own industry. The participants were convinced that they improved the viability of their business model during the focus group workshop. The following sections elaborate on how the pattern-based design addresses the three viability challenges and explain the related consequences for the business model. At the end, the adapted version of the business model is presented (cf. Fig. 4).

Tackling Viability Challenge 1 to Ensure Value Creation. The first viability challenge of the business model baseline is due to the fact that the value propositions for both customer groups, i.e. pregnant women/new mothers and health professionals, are not compelling enough to attract sufficient customers.

Regarding the first customer group (pregnant women/new mothers), the e-health business model design patterns "crowdsourcing" and "patient network" inspired the focus group to come up with the "women community" cluster that contains the ideas of integrating a community function within the app where women can exchange their pregnancy related knowledge and experience. This entails an extension of the value proposition ("communicate with other women") and of the customer relationship ("women community") for the pregnant women/new mothers segment (cf. Fig. 4). Additionally, the cluster "partner network" was built on the ideas of including pregnancy related partner offerings (e.g. from franchise companies) into the app that could be offered to the women. The ideas were developed based on the e-health business

Table 2. Relevant e-health business model design patterns, generated ideas, and their clusters

E-health business model design pattern	Idea	Cluster
Third party channel	Work with insurance companies	Go-to-market partner
Partnership for trust	Partner up with trusted insurance as well as with trusted technology company (to ensure data safety)	Go-to-market partner
Pay-per-use	Provide services on a usage basis	Go-to-market partner/Monetization of data
Secure platform	Raise confidence with trusted technology partner	Monetization of data
Healthcare data selling	Data selling to «good» parties (e.g. public health associations)	Monetization of data
Open ecosystem	Open app to partners offering pregnancy related products or services	Partner network
Franchising	Work with franchises to enlarge offering (e.g. sport franchises)	Partner network
Commission-based revenue	Expand offerings via partners for a commission per transaction	Partner network
Crowdsourcing	Provide access to the experience and knowledge of the «crowd»	Women community
Patient network	Pregnant women/new mothers community to raise attractiveness	Women community
Automation	Direct access to health professional appointments	Extended health professional services
24/7 telehealth	Offer remote consultation service	Extended health professional services

model design patterns "open ecosystem" and "franchising". This leads to an adaption of the canvas (cf. Fig. 4): The network partners (i.e. suppliers of additional offerings) are a new customer group that is provided with access to potential customers via the app. They can offer their products and services via an app-integrated marketplace. The women's value proposition is thereby extended as they can conveniently shop pregnancy related products and services and might profit from special offerings from the network partners. The app provider has the partner acquisition and management as well as the initial marketplace design as new key activities. Moreover, the business model has to reflect the corresponding costs.

The cluster "extended health professional services" not only addresses the value creation for pregnant women/new mothers but also for health professionals. Based on the e-health business model design patterns "automation" and "24/7 telehealth" the idea is to extend the offering of health professionals via the app by giving the women the opportunity to directly book an appointment with a health professional in their area or even offering them a remote consultation service. The focus group participants adapted

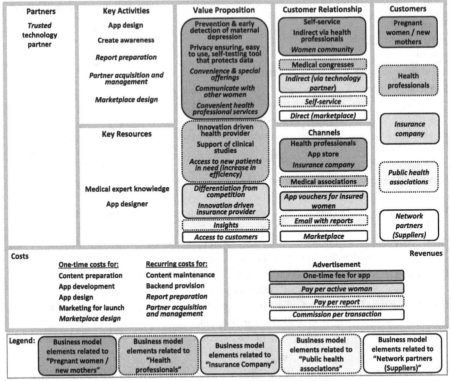

Note: Adaptions with regard to the business model baseline are in *italic*

Fig. 4. Updated business model based on e-health business model design patterns in form of the business model canvas [2]

the canvas accordingly (cf. Fig. 4): On the one hand, the value proposition of the pregnant women/new mothers segment is extended as the app offers additional services and as a consequence more value to this customer group. On the other hand, the value proposition for the health professionals is adapted as the app provides them with access to new potential customers by recommending them within the app. Since the app only refers the women to the health professionals in case they are at risk of maternal depression, the efficiency of the health professionals' offerings is increased. Thereby they can focus on the women in need of a health professional and provide them with their expertise regarding the diagnosis and treatment of maternal depression.

By enhancing the value propositions for pregnant women/new mothers as well as for health professionals, the viability challenge regarding value creation is addressed.

Addressing Viability Challenge 2 to Enhance Value Delivery. The second viability challenge for the business model baseline roots in the channel to the pregnant women/new mothers as health professionals will probably not suffice to win enough women as customers.

The "go-to-market partner" cluster includes the idea to work with a trusted insurance company. Here, the focus group judged the partnership with a trusted technology provider as a prerequisite of convincing an insurance company to provide the app to their female customers. The participants came up with their ideas based on the e-health business model design patterns "third party channel" and "partnership for trust". Integrating these ideas into the business model baseline leads to the following changes in the canvas (cf. Fig. 4): The insurance company is an additional customer group which is offered the value proposition to appear as an innovation driven company to differentiate itself from the competition. The insurance company can provide the app to its female customers by handing out "app-vouchers". Since the envisioned technology partner already has established relationships with all big insurance companies, the app provider will only have an indirect relationship (via the technology provider) with the insurance company.

The cluster tackles the viability challenge regarding value delivery as the insurance company is an additional channel to the women.

Acting on Viability Challenge 3 to Improve Value Capture. The last challenge with regard to the business model baseline is that advertising and the one-time fee for the app will not generate enough revenues for the app provider.

The above mentioned clusters "partner network" and "go-to-market partner" not only tackle the viability challenges 1 and 2, but also address viability challenge 3 that focuses on value capture. The "partner network" cluster includes the idea of opening the app to suppliers who can offer their products and services via an app-integrated marketplace and pay in turn a commission per transaction once a pregnant women/new mother buys one of their products or services. This idea is based on the e-health business model design pattern "commission-based revenue". The "go-to-market partner" cluster contains the concept of an insurance company that provides the app to their female customers whereas the insurance company pays the app provider depending on the number of insured women using the app. This thought was triggered by the e-health business model design pattern "pay-per-use". The two clusters lead to the inclusion of the two new revenue streams "commission per transaction" and "pay per active woman" (cf. Fig. 4).

In addition to the aforementioned clusters, the "monetization of data" cluster also aims at tackling viability challenge 3. Based on the e-health business model design patterns "pay-per-use", "secure platform", and "healthcare data selling", the focus group participants came up with the idea to sell the generated data to "good" (i.e. non-commercial) parties on a usage basis. Again, a trusted technology partner is a prerequisite as he signals trustworthiness by ensuring data safety. As a consequence, the canvas is updated in the following way (cf. Fig. 4): Public health associations (as non-commercial parties) are a new customer group that can buy insights regarding maternal depression in form of reports. These reports can be ordered via the app and are send out to the associations via email after a fee per report is received. Hence, an additional key activity is the report preparation which also leads to additional costs.

Overall, the network partners, the insurance company, as well as the public health associations are new revenue sources for the app provider and, under the assumption that the involved costs are lower than the respective revenues, tackle the viability challenge regarding value capture.

5   Conclusion

The goal of this paper was to understand how exactly e-health business model design patterns support the design of viable business models for e-health services. For this purpose, a focus group workshop was conducted that dealt with the business model design for a specific e-health service – an e-mental health app for maternal depression. When comparing the pattern-based business model design with their original business model idea for the e-mental health app, the focus group participants did not only appreciate the pattern-based business model design in general, but were also convinced that the viability of their business model could be enhanced in several ways:

First, regarding value creation for the different customer segments, the updated business model provides a more compelling value proposition for both women and health professionals. Thereby the probability that these customer segments are convinced of the app's value is increased which in turn leads to an increase in the related viability.

Second, by including an insurance company as an additional channel, the value can be delivered to more women compared to before where the app provider would have been solely dependent on health professionals recommending the app to their patients.

Third, the value capture could be significantly improved by having more revenue sources than before. By having the insurance company, health associations, and the network partners generating revenues, the viability could be enhanced. Moreover, since more women are convinced of the app due to a better value creation, the revenue mechanisms from the baseline (advertising and one-time fee for app) are leading to more revenues as well.

Overall, the focus group workshop revealed that the e-health business model design patterns helped to sensitize the participants to the viability aspects of the business model and to focus on the value creation, delivery, and capture for all parties involved.

Despite the focus group's appraisal of the pattern-based business model design, one has to be aware of the fact that their adaptions to the business model and the expected impacts on viability are based on certain assumptions. The participants judged their assumptions to be realistic, however, a next step would be to further validate them. For example, one assumption of the updated business model is, that an insurance company would like to offer digital services to its customers and that it would be willing to cooperate with the app provider. Another assumption is that pregnant women are ready for e-health services and would be willing to use an e-mental health app for maternal depression. This assumption is consistent with a study performed by McKinsey which revealed that more than 75 % of all respondents (across all countries) would appreciate e-health services [1]. Another study showed that women would like to use an e-mental health app regarding maternal depression [7]. Additionally, the focus group participants assumed that pregnant women/new mothers would be interested in discussing their health status with other women. As other patient networks already proved that people like to share their conditions with others [30], this assumption could be valid.

In addition to the not yet verified assumptions, this paper is not without limitations. We applied the e-health business model design patterns in only one focus group to understand how the pattern-based business model design works and if the business' viability can be supported. Moreover, even though the focus group participants judged the viability to be enhanced after applying the design patterns, there is no guarantee for market success as the business model has to be implemented yet. However, the focus group judged the pattern-based business model design as means to a viable business model and therefore as a precondition for market success.

Despite the limitations, the paper contributes to research and practice: On the one hand, it enhances extant research on business models as it gives deeper insights on the support of design patterns during business model design. On the other hand, practitioners get an idea how e-health business model design patterns can facilitate the design of viable business models for their own e-health services.

Future research should conduct further pattern-based business model designs and should also analyze the businesses over a longer period of time to see if a pattern-based business model design leads to market success or not.

References

1. Biesdorf, S., Niedermann, F.: Healthcare's Digital Future. McKinsey & Company, New York (2014)
2. Osterwalder, A., Pigneur, Y.: Business Model Generation. Wiley, Hoboken (2010)
3. Mettler, T., Eurich, M.: A design pattern-based approach for analyzing e-health business models. Health Policy Technol. 1, 77–85 (2012)
4. Spil, T., Kijl, B.: E-health business models: from pilot project to successful deployment. IBIMA Bus. Rev. 1, 55–66 (2009)
5. Sprenger, M., Mettler, T.: On the utility of e-health business model design patterns. In: European Conference on Information Systems (ECIS), Istanbul, Turkey (2016, forthcoming)
6. Danaher, B.G., Milgrom, J., Seeley, J.R., Stuart, S., Schembri, C., Tyler, M.S., Ericksen, J., Lester, W., Gemmill, A.W., Lewinsohn, P.: Web-based intervention for postpartum depression: formative research and design of the MomMoodBooster program. JMIR Res. Protoc. 1, e18 (2012)
7. Haga, S.M., Drozd, F., Brendryen, H., Slinning, K.: Mamma mia: a feasibility study of a web-based intervention to reduce the risk of postpartum depression and enhance subjective well-being. JMIR Res. Protoc. 2, e29 (2013)
8. Osma, J., Plaza, I., Crespo, E., Medrano, C., Serrano, R.: Proposal of use of smartphones to evaluate and diagnose depression and anxiety symptoms during pregnancy and after birth. In: IEEE-EMBS International Conference on Biomedical and Health Informatics (BHI), pp. 547–550. IEEE, Valencia (2014)
9. Gaynes, B.N., Gavin, N., Meltzer-Brody, S., Lohr, K.N., Swinson, T., Gartlehner, G., Brody, S., Miller, W.C.: Perinatal Depression: Prevalence, Screening Accuracy, and Screening Outcomes. Agency for Healthcare Research and Quality, Rockville (2005)
10. Seeley, S., Murray, L., Cooper, P.J.: The outcome for mothers and babies of health visitor intervention. Health Visit. 69, 135–138 (1996)
11. Petrou, S., Cooper, P., Murray, L., Davidson, L.L.: Economic costs of post-natal depression in a high-risk British cohort. Br. J. Psychiatry 181, 505–512 (2002)
12. Field, T.: Prenatal depression effects on early development: a review. Infant Behav. Dev. 34, 1–14 (2011)

13. Grote, V., Vik, T., von Kries, R., Luque, V., Socha, J., Verduci, E., Carlier, C., Koletzko, B., The European Childhood Obesity Trial Study Group: Maternal postnatal depression and child growth: a European cohort study. BMC Pediatr. **10**, 14 (2010)

14. Hewitt, C.E., Gilbody, S.M.: Is it clinically and cost effective to screen for postnatal depression: a systematic review of controlled clinical trials and economic evidence. BJOG: Int. J. Obstet. Gynaecol. **116**, 1019–1027 (2009)

15. Santoro, K., Peabody, H.: Identifying and treating maternal depression: strategies and considerations for health plans. National Institute for Health Care Management (2010)

16. O'Mahen, H.A., Woodford, J., McGinley, J., Warren, F.C., Richards, D.A., Lynch, T.R., Taylor, R.S.: Internet-based behavioral activation - treatment for postnatal depression (Netmums): a randomized controlled trial. J. Affect. Disord. **150**, 814–822 (2013)

17. Carlbring, P., Nilsson-Ihrfelt, E., Waara, J., Kollenstam, C., Buhrman, M., Kaldo, V., Söderberg, M., Ekselius, L., Andersson, G.: Treatment of panic disorder: live therapy vs. self-help via the Internet. Behav. Res. Ther. **43**, 1321–1333 (2005)

18. Grime, P.R.: Computerized cognitive behavioural therapy at work: a randomized controlled trial in employees with recent stress-related absenteeism. Occup. Med. **54**, 353–359 (2004)

19. Proudfoot, J., Ryden, C., Everitt, B., Shapiro, D.A., Goldberg, D., Mann, A., Tylee, A., Marks, I., Gray, J.A.: Clinical efficacy of computerised cognitive-behavioural therapy for anxiety and depression in primary care: randomised controlled trial. Br. J. Psychiatry **185**, 46–54 (2004)

20. Warmerdam, L., Riper, H., Klein, M., van den Ven, P., Rocha, A., Ricardo Henriques, M., Tousset, E., Silva, H., Andersson, G., Cuijpers, P.: Innovative ICT solutions to improve treatment outcomes for depression: the ICT4Depression project. Stud. Health Technol. Inform. **181**, 339–343 (2012)

21. Mizik, N., Jacobson, R.: Trading off between value creation and value appropriation: the financial implications of shifts in strategic emphasis. J. Mark. **67**, 63–76 (2003)

22. Fielt, E., Huis In't Veld, R., Vollenbroek-Hutten, M.: From prototype to exploitation: mobile services for patients with chronic lower back pain. In: Bouwman, H., De Vos, H., Haaker, T. (eds.) Mobile Service Innovation and Business Models. Springer, Berlin (2008)

23. Alexander, C., Ishikawa, S., Silverstein, M., Jacobson, M., Fiksdahl-King, I., Angel, S.: A Pattern Language. Oxford University Press, New York (1977)

24. Weill, P., Vitale, M.R.: Place to Space: Migrating to eBusiness Models. Harvard Business School Press, Boston (2001)

25. Abdelkafi, N., Makhotin, S., Posselt, T.: Business model innovations for electric mobility - what can be learned from existing business model patterns. Int. J. Innov. Manag. **17**, 1340003-1–1340003-41 (2013)

26. Rappa, M.: http://digitalenterprise.org/models/models.html

27. Gassmann, O., Frankenberger, K., Csik, M.: The Business Model Navigator: 55 Models That Will Revolutionise Your Business. Pearson Education Limited, Harlow (2014)

28. Enkel, E., Mezger, F.: Imitation processes and their application for business model innovation: an explorative study. Int. J. Innov. Manag. **17**, 1340005-1–1340005-34 (2013)

29. Mantzana, V., Themistocleous, M., Irani, Z., Morabito, V.: Identifying healthcare actors involved in the adoption of information systems. Eur. J. Inf. Syst. **16**, 91–102 (2007)

30. PatientsLikeMe. http://www.patientslikeme.com/

31. Tremblay, M.C., Hevner, A.R., Berndt, D.J.: The use of focus groups in design science research. In: Hevner, A.R., Chatterjee, S. (eds.) Design Research in Information Systems: Theory and Practice, pp. 121–143. Springer, New York (2010)

32. Frank, U.: Evaluation of reference models. In: Fettke, P., Loos, P. (eds.) Reference Modeling for Business Systems Analysis, pp. 118–140. Idea Group, London (2007)

33. Doll, J., Eisert, U.: Business model development & innovation: a strategic approach to business transformation. 360° – Bus. Transform. J. **11**, 6–15 (2014)
34. D'Souza, A., Wortmann, H., Huitema, G., Velthuijsen, H.: A business model design framework for viability; a business ecosystem approach. J. Bus. Models **3**, 1–29 (2015)

National/Regional Initiatives in e-Health and Welfare

The Emperor with No Clothes – Inter-organizational ICT Cooperation Within Municipal Regions

Tomi Dahlberg[⊠] and Ari Helin

Turku School of Economics, Information Systems Science,
University of Turku, 20014 Turun Yliopisto, Finland
{tomi.dahlberg,ari.helin}@utu.fi

Abstract. ICT has a significant role in the development and production of municipal services, and in the daily work of municipal civil servants. Yet, municipalities typically develop and operate their ICT activities independently with limited ICT resources, both money and people. Limited resources are the key incentive for inter-municipal ICT cooperation. This article investigates how ICT cooperation is carried out within 20 municipal regions. As the theoretical basis we review transaction cost economics (TCE), resource based view (RBV) and the constructs of Granovetter's social network theory. They are used to outline potential economic and social benefits emerging from ICT cooperation and to describe social mechanisms that influence the realization of such benefits. Our empirical data reveals that there are distinct differences in the effectiveness of ICT activities, in the amount of ICT cooperation and in the governance of ICT cooperation. Our analysis reveals that the emperor will not have clothes – ICT cooperation benefits – unless the cooperation is properly organized and governed.

Keywords: ICT governance · ICT cooperation · Municipalities

1 Introduction

ICT is deployed extensively in the daily municipal activities. It has become an integral part of more or less all municipal activities and services. Municipal civil servants rely on ICT in their work. ICT is deployed to develop and produce social welfare, healthcare, educational, technical and all other services that a municipality provides to its citizens, companies and organizations. It is also widely acknowledged that without ICT municipalities cannot be managed since data about the various activities of municipalities are created, processed, stored and reported with ICT. Moreover, the role of ICT in the activities and services of municipalities appears to be ever increasing. Digitalization of services, digitization of printed and analog legacy materials and Internet of things, for example in infrastructure surveillance, to name just a few emerging ICT deployment areas, are examples of the ever-increasing role of ICT.

Yet, municipalities develop and operate ICT activities with limited ICT resources, both money and people. A medium-size municipality in the country of the present study (Finland) with 20 000 inhabitants, may have thousands of network-connected

© Springer International Publishing Switzerland 2016
H. Li et al. (Eds.): WIS 2016, CCIS 636, pp. 179–192, 2016.
DOI: 10.1007/978-3-319-44672-1_15

devices in offices, schools and other premises of the municipality. In addition to local device specific applications, data and networks, these devices are used to access the dozens/hundreds of information systems (IS) and applications that are used to produce various municipal services. Devices are also provide access to databases, server computers and telecommunication networks that constitute the ICT infrastructure of these IS and applications. Physically ICT infrastructure is usually placed into one or a few data centers. In a medium-size municipality the one data center could be a locked storage room. A medium-sized municipality often has two to five full-time IT specialists to take care and manage all of the above-described ICT. A smaller community with a few thousand inhabitants may have none. It appears as if municipalities had not recognized that their activities and services have become entirely IT-dependent.

The social welfare and healthcare reform in Finland constitutes the background of our research. The National Government decided in late 2015 that 18 municipal regions should have the responsibility for arranging these services from the beginning of 2019. The purpose is that the municipal regions would cooperate also in the arrangement of other municipal services, their ICT included. The data of a countrywide research [1] funded by the Ministry of Finance was made available to our research. The data includes information about Finnish municipalities, such as their size, the number of civil servants, the number of ICT personnel, and the amount of ICT spending. The data also covers the ICT cooperation within 20 municipal regions. We also use data from two other studies [2, 3]. In summary, the purpose of the present study is to investigate how Finnish municipalities cooperate or do not cooperate in ICT by analyzing data made available to us, to outline potential cooperation benefits and to depict social mechanisms that may realize or prevent cooperation and benefit achievement.

In talks, scarcity of ICT resources is regarded as an incentive for ICT cooperation between nearby municipalities. Often heard claims are that with ICT cooperation (nearby) municipalities can pool and share ICT resources, develop and produce similar ICT services together at lower costs per municipality and achieve cost savings by conducting pooled joint hardware and software purchases, which lead to lower unit prices. The transaction cost economics (TCE) theory [4] and the resource based view (RBV) [5] offer theory-based support for the benefits of ICT cooperation. The potential to achieve benefits is what we mean with the emperor having new clothes. On the other hand, municipalities differ in size as well as in economic and industrial structures. These differences and municipalities' constitution-based legal independence have to be taken into account should one wish to engage municipalities into ICT cooperation. The constructs of Granovetter's social network theory [6–9] offer theoretical explanations, why efforts to establish ICT cooperation may not lead to the emperor actually getting new clothes.

The objective of our research is demonstrate that the TCE theory and the RBV theory together with the constructs of Granovetter's social network theory provide solid theoretical basis for understanding the potential benefits of ICT cooperation and social mechanisms that influence the ability to achieve such cooperation within the investigated municipal regions. Our second objective is to understand what are the current status in the effectiveness of ICT activities within the investigated municipalities as well as the status of the inter-organizational ICT cooperation in terms of the volume and depth of ICT governance. To achieve these objectives we raise the following research questions:

1. How do Finnish municipalities cooperate currently in ICT and how is ICT cooperation organized?
2. What ICT cooperation benefits and possibilities have Finnish municipalities identified and do the identified benefits and opportunities act as antecedents for willingness to engage into (formal) inter-organizational IT governance arrangements?
3. If ICT cooperation between Finnish municipalities is organized on the basis of an (formal) inter-organizational IT governance arrangement or alternatively on the basis of an ad-hoc arrangements, do those alternatives impact the achievement of ICT cooperation and cooperation benefits?

The rest of the paper is organized as follows. As the theoretical background of our study we depict what potential benefits the TCE theory and the RBV approach propose for inter-organizational ICT cooperation. In the theoretical background section we also describe how social network theory constructs impact the achievement of ICT cooperation benefits. We then explicate our research methods and the characteristics of the empirical data sets used in the data analysis. After the results of the study are disclosed we end the article with a discussion of its scientific and practical contributions and conclusions for researchers and practitioners.

2   Theoretical Background

2.1   TCE and RBV and the Economic Benefits of ICT Cooperation

The transaction cost economics theory calls economic transactions inside a single organisation (make) as vertical governance, and transactions between a buyer and a seller as market governance (buy). Market governance transactions with one principal and several subcontractors establish a hierarchical network and transaction in alliances and networks constitute relational, also known as networked, governance [4, 10]. Inter-organizational cooperation between municipalities falls into the category of relational governance. The rationale of the TCE theory is that an organization (municipality) should rely on that form of governance that provides lowest transaction costs [11]. TCE proposes that market governance, such as reliance on municipal ICT cooperation, is most viable when assets are non-idiosyncratic, the use of assets is voluminous or very small and the uncertainty related to asset usage outcomes can be lowered [4, 10, 11]. The simplistic interpretation is that municipalities should conduct ICT cooperation in issues that are similar to all of them and where they are able to reduce risks (uncertainty) by cooperation. Cumulatively, TCE research has identified dozens of cost-inclusive and cost-exclusive metrics [11] to measure the cost savings of asset specificity, uncertainty and frequency.

The resource based view approach considers each organisation as unique [5, 12]. An organisation (municipality) is equivalent to the broad set of resources that the organization owns at least (semi-) permanently [13]. These resources as a whole - especially those resources that are imperfectly mobile, imitable, and substitutable - define the value creation potential of the organisation. Thus the RBV approach focuses on how to add value when TCE focuses on how to save costs. The idea of RBV is that and organization (municipality) should rely inter-organisational (municipal ICT) cooperation if that provides more value to cooperation participants through pooling, aggregating, sharing, and exchanging their unique resources. For cooperation to happen it is also necessary that added win-win value couldn't be achieved easily in other ways. The conclusion is similar to the TCE theory. Municipalities should cooperate in ICT, when they face similar issues and when cooperation provides measurable value to all of them. Value increases detected in RBV research are classified into the following categories: conserve resources, share risks, obtain information, access complementary resources, reduce product development costs, improve technological capabilities, and enhance reliability [14].

2.2   Constructs of Granovetter's Social Network Theory

We use the constructs of Granovetter's social network theory to understand why the ICT cooperation potential between municipalities happens or is prevented. Other theories about social structures and mechanisms fall outside the scope of our study. For example, we do not use the social network analysis approach [15], in which connections between the individuals of a network are crafted into a matrix and analyzed to detect such properties of the network as cohesion, centralicity and power.

Social networks affect the flow and the quality of information between individuals as well as the trust that other individuals do the right things in a social network [9]. Also the TCE theory emphasizes the importance of trust as a property of relational governance. Existence of trust is seen to impact positively both the creation and the performance (of municipal ICT) cooperation [11]. The same view is present also in RBV. Prior studies have reported that the lack of trust reduces willingness to cooperate. These findings justify the inclusion of social network theory constructs into the theoretical background of our research, since social network theory constructs explain the existence or lack of trust [13].

The density of a network is the first relevant construct [9]. The denser a network the easier it is to enforce norms to the network, i.e. agree the objectives of municipal ICT cooperation. The practical interpretation is that should there be close long-term and focused cooperation between municipalities in ICT, then more results could be expected. The research [2] to be reviewed in Sect. 2.3 discovered that the past history of municipal ICT cooperation in Finland could best be described with words ad-hoc cooperation and non-existence of organized ICT governance. The larger a network the lower the network density, because people have cognitive, emotional, spatial and temporal limits on how many social ties they are able sustain [7, 9]. Sparse networks (low density) may explain why municipal ICT professionals have not cooperated even if that would help them professionally.

Any organization consists of several different social groups, to which individuals are connected with strong ties. Through weak ties, the second relevant concept, individuals are connected to groups outside their own social circle. Weak ties between organizational groups provide access to information and resources beyond those available in one's own social circle [6, 7]. The practical meaning is that should there be long-term and frequent communicative connections between the ICT experts of municipalities, then that would have positive impacts on ICT cooperation and benefits achievement.

The structural hole construct [9] extends the weak ties construct. The structural hole construct proposes that is even more important that different parts of a network are connected than what is the nature of ties. The practical implication is that the lack of long-term and frequent connections between municipal ICT experts will influence ICT cooperation negatively. Social embeddedness, also called the interpenetration of economic and non-economic action [8], is the final relevant construct discussed. The performance objectives of ICT cooperation typically have strong economic incentives. Ability to save IT cost or to develop ICT services efficiently are examples of such incentives. On the other hand, social connections are largely non-economic [9]. The practical implication is that the outcomes of ICT cooperation between municipalities should not based only on economic objectives and metrics but also on social metrics, for example on the development of professional competencies through knowledge and best practices sharing.

2.3 Previous Research on the ICT Cooperation Between Finnish Municipalities

Our previous study [2] described, how inter-organizational IT governance was established between five hospital districts, 68 municipalities and 33 healthcare centers in Finland during the years 2013–2014. In the present study we depict only that part of article, which addresses the expected benefits of systematic inter-organizational IT governance. The establishment of the inter-organizational ICT governance arrangement [2] resulted in the discovery of 13 perceived ICT cooperation benefits shown in Table 1. Fifteen individuals representing the mentioned hospital districts, municipalities and healthcare centers defined the benefit statements of Table 1 in cooperation. They also crafted and implemented the inter-organizational ICT cooperation governance arrangement to realize the identified cooperation benefits [2]. To validate the identified cooperation benefits a national survey was arranged to do that. An invitation to participate to the survey was sent to 240 social welfare and healthcare specialists and 68 valid responses were received. A respondent was requested to evaluate each of the 13 benefit statements. A 7-step Likert scale from fully agree with the benefit statement (response value = 7) to fully disagree with the benefit statement (response value = 1) was used in the survey. Table 1 shows the proportion of responses that either fully (response value = 7) or strongly agreed (response value = 6) with a benefit statement. Table 1 also shows the connection of the benefit statements to TCE and RBV, which were used to craft the benefit statements (see [2]).

On the basis of the survey it is evident that the similar ICT cooperation benefits are available to other municipalities.

Table 1. Perceived benefits of ICT cooperation between municipalities

Perceived benefit of inter-organizational ICT cooperation in municipal healthcare and social welfare ICT with links to theoretical TCE and RBV constructs [2]	Proportion of strongly agree (n = 68)
Avoid the development of overlapping and difficult to integrate IT services (RBV- conserve resources; TCE – asset specificity)	86.8 %
Increase the interoperability of patient/customer information systems and data storages (RBV – obtain resources; TCE – asset specificity	86.8 %
Create enterprise architectures (RBV – substitute; TCE reduce uncertainty)	79.4 %
(Co-)source IT-services cost-efficiently and effectively (RBV – create difficult to imitable resources; TCE – reduce cost of non-specific assets)	75.0 %
Implement national level healthcare and social welfare IT services efficiently and effectively (RBV – access to resources that are difficult to imitate; TCE – acquire assets that are specific at low costs)	75.0 %
Ensure ability to participate to the national level development of healthcare and social welfare services (RBV and TCE as above)	72.1 %
Use IT resources and assets efficiently and effectively (RBV and TCE – all key constructs of both theories)	70.6 %
Ensure access to specialized capabilities and competencies everywhere in the area (RBV – ensure access to rare resources; TCE – ensure availability of idiosyncratic assets)	69.1 %
Ensure availability of equal healthcare and social welfare services everywhere in the area (RBV and TCE as above)	67.6 %
Tighter cooperation on national level (RBV – substitute resources; TCE reduce uncertainty)	89.7 %
Tighter cooperation on regional level (this area) (RBV – share unique resources; TCE – reduce the impact of idiosyncratic assets)	85.3 %
The creation of jointly agreed data models and sticking to them (RBV – value through pooling; TCE reduce uncertainty)	85.3 %
Tighter cooperation between healthcare and social welfare (RBV – create substitute value; TCE – reduce behavioral uncertainty)	79.4 %

3   Methodology

Our empirical research material consists of information from three case studies. One of them was a large research that covered most Finnish municipalities and the two other researches were regional case studies in Northern Finland and Northern Ostrobothnia. The empirical research material of [1] was collected in a research conducted by the Ministry of Finance. The purpose of their research was to support municipalities in their efforts to restructure municipal services. The name of the research was "ICT change support program" (ICT-muutostukiohjelma). Cumulatively 144 out of the country's 317 municipalities participated into this research during 20.2.2014–31.12.2015. Municipalities were grouped into 20 regional areas. Each group produced a report on ICT within its regional area. We analyze the data as secondary data.

The empirical research material includes information about participating municipalities. The following background information was collected: name of the municipality, number of ICT personnel, the proportion of ICT personnel from the total number of municipal civil servants, the format of municipal cooperation and the names of participating municipalities in the regional cooperation. From data we collected comments and observations on the following topics: the format of municipal cooperation, the core of ICT cooperation, other cooperation entities, and ownership in Kuntien Tiera Oy. Kuntien Tiera is a nationwide ICT service company owned by municipalities. To verify the size of population in each municipality, we used data provided by the Finnish Population Register Center (Väestörekisterikeskus) [16].

Research material contained 20 reports on regional municipal areas as participating municipalities were divided into 20 regional areas. Each regional area was considered as the candidate of municipal mergers between the municipalities of the area. The eleven largest urban areas of the country were included. The number of municipalities varied between 3 and 17. Three regional reports followed a different approach to the other 17 regional reports. These three reports focused only to specific details or future scenarios. Due to the inconsistencies between these three and the other 17 reports we excluded them from data analysis.

We read all remaining 17 reports and classified information in them as explained earlier. Still, it is necessary to note that the approaches and the content of the 17 reports are not fully compatible and comparable. The reason is that consultants wrote the regional reports. This resulted in report inconsistencies. Information collected and analyzed in the present study was, however, mostly available.

As mentioned, in addition to the research material made available to us by the Ministry of Finance, we used data from two other case studies. Mentioned studies reported well-organized ICT cooperation and/or governance arrangements between municipalities.

4 Results

4.1 ICT Cooperation Between Municipalities (Research Question 1)

The size of the population in municipalities ranged from 755 inhabitants to 623732 inhabitants [16]. Due to the large variability in the population size of municipalities, also the number of civil servants and the number of ICT personnel employed by municipalities varied a lot. The number of ICT personnel varied from zero to 483 employees. Statistics on the number of ICT personnel was compiled into Table 2.

Most small municipalities had no or very few ICT experts. Table 2 indicates that 27 municipalities, almost 20 % of the 139 municipalities, did not have any ICT expert. In addition to that, 26 municipalities reported that the number of ICT personnel is less than one. In these 26 municipalities, other municipal civil servants carried out ICT tasks in addition to their other duties or a part-time person had been recruited. Twenty-nine plus five municipalities did not report the number of ICT personnel or the numbers of ICT personnel were summed up to the total number of ICT personnel in a regional report. The comparison between the number of ICT personnel and the number

Table 2. The number of ICT personnel

Number of ICT personnel	Number of municipalities
0	27
≤ 1	26
1 < 2	8
2–9	35
10–49	7
50–99	4
100+	3
No information/unknown	29/5

of municipal civil servants revealed that the ratio varied from 0 and 2.4 %. Two point four percent was detected in a very small municipality, where one ICT professional was sufficient to produce this ratio. The ratio of ICT personnel to total number of municipal civil servants was above one percent in only 9 municipalities. Seven of them had less than 3300 inhabitants. The other two municipalities are among the five largest municipalities in Finland. There appears to be no correlations between the size of the population in a municipality, the number of municipal civil servants and the number of ICT personnel. We did, however, not perform any statistical tests.

Data on the depth of ICT cooperation was also compiled from the regional reports. Table 3 was created by classifying the comments and feelings of persons that participated into the research conducted by the ministry. Deep or fairly deep cooperation can be interpreted as regular, (semi-)official cooperation. Limited and very limited cooperation means ad-hoc cooperation with information sharing from time to time. As Table 3 illustrates limited ICT cooperation between nearby municipalities was more common than deep cooperation.

Table 3. Municipal ICT cooperation

Depth of ICT cooperation	Number of regions (n = 20)
Deep	3
Fairly deep	3
Some	4
Limited	9
Very limited	1

ICT cooperation between municipalities is most often conducted in healthcare and social welfare ICT. The insufficiency of ICT resources, both financial and human, was the most often mentioned reason for the lack of ICT cooperation. The number of ICT personnel is so small that it has been necessary to allocate all resources to keep the operational ICT systems running. That left no ICT experts to municipal ICT cooperation or to joint development of ICT services. The infamous double bind (catch-22) concept could be used to describe this finding.

Even though ICT cooperation between municipalities appears to be limited in Finland, largely due to lack of time available, municipalities have cooperated since times immemorial. The Constitution and other laws allow Finnish municipalities to organize their service production in multiple ways. It often makes sense to join forces and organize the production of services together. The history of municipalities has witnessed several forms of cooperation. One common practice is to create a joint authority to take care of a specific service area. There were 136 such joint authorities in 2012 in Finland. A joint authority is an independent public entity, which operates under the legislation governing the activities of municipalities and is "owned" by the cooperating municipalities. Healthcare districts are the largest joint authorities. Municipalities may also jointly hire civil servants to produce selected services. Especially smaller municipalities use this option. Municipalities have established jointly owned enterprises to take care of waste management, business services and tourism. They have also signed cooperation agreements to execute such services as water management, rescue services, educational activities and the inspection of buildings. Municipalities may even buy services from other municipalities, should that make economic sense. Regional cooperation between municipalities within a specific service area has traditionally been associated with industrial policy and lobbying [17]. We find it surprising that municipalities have established only a few joint ICT authorities, such as Kuntien Tiera, to ensure that benefits described in Sect. 2.3 could be obtained. We feel that sparse networks with structural holes and lack of weak tie, which make true cooperation possible, are important determinants for this phenomenon.

4.2 Perceived Benefits and Potential of Municipal ICT Cooperation (Research Question 2)

Our data analysis showed that ICT cooperation between municipalities is most often loose and based on ad-hoc arrangements. Sixteen out of the 17 regional areas were willing to consider ideas on how to organize municipal ICT cooperation better, as Table 4 indicates.

Fourteen of the regional reports included data on more detailed development ideas for municipal ICT cooperation. We have compiled these results into Table 5. The titles of the development ideas are shown as expressed by the participants.

Table 4. Development ideas to improve municipal ICT cooperation

Development idea to improve ICT cooperation	Number of areas willing to consider the idea (n = 16)
Data center cooperation or server cooperation	9
Joint procurement	7
Shared development, national cooperation	6
Development cooperation with a municipal group	4
Buying shares in an existing company or organization	3
Establishment of a new, shared service ICT company	2
Discussion forums	2
Mergers between municipalities	1

Table 5. Detailed development ideas for municipal ICT cooperation

Detailed development idea	Number of responses (n = 14)
Integration of IS	11
Centralization of ICT infrastructure	8
Development of (regional) digital services	7
Renewal of ICT model	6
Process harmonization	5
Organization and development of ICT expert resources	4
Development of operating models	4
Closer cooperation between business and ICT management	3
Harmonization of contracts	3
Joint development of ICT governance models	3
Joint Strategic Management of ICT	3
Shared ICT service operations, ICT service strategy	3
Development of a common guidance	2
Development of ICT awareness among municipal managers	2
Development of special knowledge	1
Development of ICT design activities	1
Development of ICT design methods	1
Shared management of ICT governance models	1
Identification of ICT services for core business	1

Seven regional reports contained data on regional development projects. We compiled that data into Table 6. Project names are labeled as expressed by the participants.

In the regional reports, several municipalities described their ICT cooperation with other municipalities as loose and based on ad-hoc arrangements. Civil servants and ICT professionals meet and diverse topics are discussed. Yet, since no formal ICT governance arrangement has been established topics remain open. ICT cooperation regarding more detailed development ideas is also described as loose and ad-hoc. Our previous study [2] described a similar situation prior the establishment of a formal ICT governance arrangement. Our conclusion is that the establishment of inter-organizational ICT governance receives little support among the ICT professionals working in municipalities, as the necessary social network structures do not exist. ICT professionals have proposed many ideas on how to improve current ICT cooperation between municipalities but are reluctant to improve ICT governance as a primary short-term activity since there is no trust. On the other hand, several cooperation benefits were mentioned as Tables 4, 5 and 6 showed. Also several detailed ICT cooperation development ideas, such as the renewal of ICT model, development of operational models, joint development of ICT governance models, and joint strategic management of ICT. At least some municipal ICT experts recognize the need to establish better-organized ICT cooperation and governance arrangements. We conclude that the necessary activities is to build necessary social ties to facilitate trust building prior to or in the connection of more formal ICT cooperation arrangements.

Table 6. Local development projects

Local development project	Number of times mentioned (n = 7)
Electronic transactions and archiving	7
Electronic meeting practices	5
Master data management	4
Learning solutions	4
Document management, records management and archiving	4
Data management, knowledge management	4
Definition of processes	3
Organization and user rights management	3
Integration principles of systems, data and the solutions	3
Project portfolio and project management	3
ICT model	3
Service portfolio definition	2
Intranet systems development and integration	2
User support, desktop support	2
Knowledge management	2
Reporting principles	1
Project work method	1

The regional reports reveal that among the 20 regions ICT governance is advanced in one particular region. That region consists of one major city and 8 smaller municipalities surrounding the city. These 8 municipalities have established an alliance, which manages and develops ICT for all municipalities and manages the relationship to the major city. Municipalities have recruited a shared ICT Director. The ICT director represents participating municipalities in relevant forums and the salary is divided between the municipalities. The ICT governance arrangement aligns ICT management activities with the major city and the municipalities. Two ICT executive committees have been organized, one with the major city and another with the surrounding municipalities. The major city and the municipalities still have their own projects, working groups and meetings to make their own decisions. Ability to provide alternatives and mobility to customers with shared regional learning environments, regional public transportation and well-functioning administration are some of main objectives in the regional inter-organizational ICT cooperation Both the major city and the surrounding municipalities have been pleased with the ICT governance arrangements and the progress of municipal ICT cooperation.

In two other cases, ICT governance was an explicitly stated objective. The project conduced in northern Finland expressed: "the purpose of the agreement and the specified ICT cooperation is to sharpen and clarify ICT steering, and to thus enhance the service capability and productivity of ICT" [3]. Participating municipalities have agreed shared objectives and development areas for municipal inter-organizational ICT cooperation. Municipalities have also signed cooperation and project agreements, and have started the implementation of shared projects. According to data available ICT cooperation has progressed well based on the inter-organizational ICT governance. The data of the other case study [2] provides similar results.

4.3   The Impacts of Inter-organizational IT Governance (Research Question 3)

On the basis of the research funded by the Ministry of Finance and two separate case studies, well organized ICT cooperation and governance has indisputable positive impacts on ICT activities and services. Business and ICT professional are more satisfied. ICT activities and services have clearer objectives. The establishment of well-organized ICT cooperation and governance arrangement also helps to solve challenges that limited ICT resources create in municipalities. Why have ICT professionals in municipalities not used these opportunities? ICT professionals appear not have the time. Yet, we propose that the real reason is also related to the deficiencies in inter-municipal social networks. Social networks are needed to build trust and to share information between municipalities with different sizes and priorities.

We summarize our results to the three research questions with the metaphor we used in the title of this article. The emperor (municipalities) would like to have new clothes (improve ICT cooperation to receive benefits). Only after the establishment of well-organized ICT cooperation governance, are they able to realistically to get new clothes (benefits from ICT cooperation based on concrete development ideas and objectives).

5   Discussion and Conclusions

The purpose of the present study was to investigate how Finnish municipalities cooperate in ICT by analyzing secondary data and to explain connections between empirical findings and theoretical constructs. Data analyzed indicated that Finnish municipalities run their ICT largely alone without well-organized ICT cooperation. The Constitution of Finland [18] stipulates that municipalities are independent self-governed entities. Thus they have the right to do so. On the other hand, there are several reasons for municipal ICT cooperation between (nearby) municipalities with significant economic and other benefits. The size of a municipality is often small in comparison to ICT investment and development needs, ICT resources and their availability to a municipality is limited, the development of municipal services are increasingly ICT-enabled and various reforms of municipal services require more investments into ICT and digitalization. These are strong drivers for cooperation between municipalities.

We discovered that municipalities consider ICT cooperation between municipalities important. Expectations and perceptions about the potential benefits of ICT cooperation are both economic and non-economic. The transaction cost economics theory and the resource based view approach are able to describe expected economic benefits. Yet, active cooperation among municipalities was detected to be relatively rare as municipalities struggled with their limited ICT resources to keep their operative ICT running.

This situation leaves little if any room for ICT development and ICT cooperation with nearby municipalities. We discovered three regions where ICT cooperation between municipalities was more advanced. Each of these three regions had a slightly different approach to ICT cooperation. One region had a shared ICT management function, another region a jointly owned ICT service company and the third area inter-organizational ICT governance arrangement. Despite of the differences these three regions perceived that ICT cooperation progressed and delivered benefits due to clearly organized ICT cooperation and accountabilities. The contrast to regions with little ICT cooperation was clear. Municipalities in these regions were dissatisfied with ICT cooperation. Due to the lack of cooperation ideas on how to improve their situation were not implemented. The constructs of the social network theory, especially network density, weak ties and structural holes explain differences between the perceptions of those who cooperate actively and those who do not. These are our responses to the three research questions and the objectives stated in the beginning of our article.

Our study was conducted in one country and caution with the results of our research is necessary. The empirical data contained only few cases with active ICT cooperation between Finnish municipalities, which is another limitation. Evidence to properly analyze well-organized and disorganized ICT cooperation is limited. These limitations can hopefully be removed with future studies by conducting similar studies in other countries and by continuing data collection on ICT cooperation outside of the 20 municipal regions and the two cases. In addition to academic contributions such research could also have practical value. Inter-organizational ICT cooperation between municipalities starts often gradually. ICT professionals want to discuss professional matters with their colleagues and share experiences. That may lead to better-organized cooperation and, as our study shows, to more benefits when ICT cooperation is well organized. We once more underline the role of social ties.

Our advice to researchers is to especially investigate the social mechanism of inter-organizational ICT governance. Our advice to practitioners is to take the time and effort to establish a well-organized arrangement for ICT cooperation and governance. In doing so they should consider how to secure strong flow of economic and social information between the cooperating municipalities and persons.

References

1. Ministry of Finance. https://wiki.julkict.fi/julkict/Kuntauudistus/ict-muutostuki-1. Accessed 30 Nov 2015
2. Dahlberg, T.: The creation of inter-organisational IT governance for social welfare and healthcare IT – lessons from a case study. IJNVO 16(1), 38–71 (2016)
3. ePS2, Tietohallintoyhteistyö, Kari Hyvönen, presentation material, 4 November 2015
4. Williamson, O.E.: Markets and Hierarchies: Analysis and Antitrust Implications – A Study in the Economics of Internal Organization. The Free Press, New York (1975)
5. Wiengarten, F., Humphreys, P., Cao, G., McHugh, M.: Exploring the important role of organizational factors in IT business value: taking a contingency perspective on the resource-based view. Int. J. Manag. Rev. 15(1), 30–46 (2013)
6. Granovetter, M.S.: The strength of weak ties. Am. J. Sociol. 78(6), 1360–1380 (1973)

7. Granovetter, M.S.: The strength of weak ties: a network theory revisited. Sociol. Theory **1**, 201–233 (1983)
8. Granovetter, M.S.: Economic action and social structure: the problem of embeddedness. Am. J. Sociol. **91**(3), 481–510 (1985)
9. Granovetter, M.S.: The impact of social structure on economic outcomes. J. Econ. Perspect. **19**(1), 33–50 (2005)
10. Williamson, O.E.: The Economic Institutions of Capitalism. The Free Press, New York (1985)
11. Geyskens, I., Steenkamp, J.-B.E.M., Kumar, N.: Make, buy, or ally: a transaction cost theory meta-analysis. Acad. Manag. J. **49**(3), 519–543 (2006)
12. Barney, J.: Firm resources and sustained competitive advantage. J. Manag. **17**(1), 99–120 (1991)
13. Das, T.K., Teng, B.-S.: A resource-based theory of strategic alliances. J. Manag. **26**(1), 31–61 (2000)
14. Park, N.K., Mezias, J.M., Song, J.: A resource-based view of strategic alliances and firm value in the electronic marketplace. J. Manag. **30**(1), 7–27 (2004)
15. Scott, J.: Social Network Analysis, 3rd edn. Sage Publications Ltd., London (2013)
16. Väestörekisterikeskus. http://vrk.fi/default.aspx?docid=8816&site=3&id=0. Accessed 17 Mar 2016
17. Kuntaliitto. http://www.kunnat.net/fi/kunnat/toiminta/yhteistoiminta/Sivut/default.aspx. Accessed 24 Mar 2016
18. The Constitution of Finland, Ministry of Justice, Finland. http://www.finlex.fi/en/laki/kaannokset/1999/en19990731.pdf. Accessed 21 Mar 2016

Framework-Based ICT Governance and Survey in Northern Savonia

Virpi Hotti[1(⊠)] and Heikki Meriläinen[2]

[1] School of Computing, University of Eastern Finland, Kuopio, Finland
Virpi.Hotti@uef.fi
[2] Istekki Oy, Kuopio, Finland
hmerilai@gmail.com

Abstract. The ICT governance function has three main tasks: evaluate, direct and monitor (EDM). We focused on four frameworks (ISO/IEC 38500, COBIT, ITIL and ICT Standard for Management) and analyzed on how the governance function is covered on them. The international standard for the corporate governance of information technology (ISO/IEC 38500) maps EDM to the principles, the Control Objectives for Information and Related Technology (COBIT) divides EDM into the processes and the IT infrastructure Library (ITIL) illustrates EDM as the activities. The ICT Standard for Management describes EDM implicitly. Based on the frameworks we designed the survey for assessing the ICT governance. The survey was sent to 331 respondents in Northern Savonia, of which 136 responded and 28 of these respondents were the members of the executive board. In this paper, we present the responses and analyze them by means of data analytics.

Keywords: ICT governance · ISO/IEC 38500 · COBIT · ITIL · ICT Standard for Management

1 Introduction

In generally, the governance function is responsible for the preparation of the strategy based on the enterprise architecture (Fig. 1). The enterprise architecture is a strategic management tool that helps steer the development of ICT and business processes towards a defined goal. The enterprise architecture describes how the elements of an organization (i.e., organization units, people, operational processes, information and information systems) are connected to each other and how they work as a whole.

The Open Group Architecture Framework (TOGAF) recognizes that within the enterprise governance can consist of multiple separate domains as follows [2]:

- General governance. To ensure "that business is conducted properly" and to ensure "sustainability of an organization's strategic objectives" by "guidance and effective and equitable usage of resources".
- Architecture governance. To manage and control enterprise architectures and other architectures "at an enterprise-wide level" by the practice and orientation.

© Springer International Publishing Switzerland 2016
H. Li et al. (Eds.): WIS 2016, CCIS 636, pp. 193–206, 2016.
DOI: 10.1007/978-3-319-44672-1_16

- IT governance. To link "IT resources and information to enterprise goals and strategies" within the framework and structure. To institutionalize "best practices for planning, acquiring, implementing, and monitoring IT performance, to ensure that the enterprise's IT assets support its business objectives".

- Technology governance. To control "how an organization utilizes technology in the research, development, and production of its goods and services".

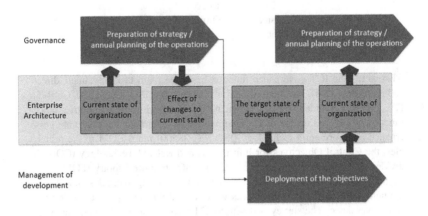

Fig. 1. Enterprise architecture and governance (adapted from [1])

TOGAF recommends the Control Objectives for Information and Related Technology (COBIT) as a good source for the IT governance and there is a mapping between earlier versions of the frameworks, i.e. between TOGAF 8.1 and COBIT 4.0. COBIT defines the governance to ensure "that stakeholder needs, conditions and options are evaluated to determine balanced, agreed-on enterprise objectives to be achieved; setting direction through prioritisation and decision making; and monitoring performance and compliance against agreed-on direction and objectives" [3]. The TOGAF content metamodel [4] the objectives, as time-bounded milestones for enterprises used to demonstrate progress towards goals tracked against measures (i.e., indicators or factors). Therefore, the governing body ("person or group of people who are accountable for the performance and conformance of the organization" [5]) has to understand performance indicators the basement of which are patterns in data and models which extrapolate into the future.

The frameworks around ICT may have parts of the governance and management functions. We found only one categorization (quality management, quality improvement, IT governance, information management and project management) around the ICT frameworks. However, the following three frameworks "focus on how to organize the IT function in terms of responsibilities, controls, organization" [1]: the Control Objectives for Information and Related Technology (COBIT), the Australian Standard for Governance of IT (AS 8015), and the Management of Risk (M_o_R).

In this paper, we focus on the frameworks around three main tasks of the ICT governance: evaluate, direct and monitor (EDM). We will illustrate how the international standard for the corporate governance of information technology (ISO/IEC 38500) maps EDM to the principles and how the Control Objectives for Information and Related Technology (COBIT) divides EDM into the governance-related processes. The Information Technology Infrastructure Library (ITIL) recognizes ISO/IEC 38500 as a standard for the ICT governance and it summarizes three main activities (evaluation, directing and monitoring). However, ITIL has five processes that are related to the ICT governance and we will illustrate those processes.

In subsequent sections, we first give an overview of the framework-based governance function. Then we describe the survey that was designed to assess the ICT governance in Northern Savonia. The material of the survey consists of 136 responses from Northern Savonia where assessing the current status of the ICT governance was made to gather information for a basis of the enterprise architecture work. It is important to assess the current status of the ICT governance, especially, to gather information for a basis of the enterprise architecture work. In this paper, we present the survey to assess the ICT governance. The questions are mainly based on the ICT Standard for Management that has been chosen by the Finnish Ministry of Finance as a governance model for ICT of the public sector. We did not have the earlier theoretical construction, which validity to be tested. Therefore, we established the questions [6] mainly based on the generic features (i.e., objectives and roles) of the overviewed frameworks. In this paper, the novelty is both in mappings between framework-based governance and the established questions, as well as, in comparison of the answers between two different sets of the respondents (i.e., the executive board and others). Finally, we discuss the results and suggestions for the future research.

2 Framework-Based Governance Function

The corporate governance usually includes principles and guidelines for using them in a practice. Some organizations talk about policies, which are collection on the related principles and guidelines for governing and managing the enterprise. Furthermore, the corporate governance may map the principles to the specific tasks.

We describe how we have formed the view of the framework-based governance. In subsequent sections, we focus on the governance function of the follows:

- the International standard for Corporate governance of information technology (ISO/IEC 38500)
- the Control Objectives for Information and Related Technology (COBIT)
- the IT infrastructure Library (ITIL)
- the ICT Standard for Management

2.1 ISO/IEC 38500

ISO/IEC 38500 is a framework, which focuses on the IT governance. The standard is intended for directors (owners, board members, directors, partners, senior executives, etc.) for evaluating, directing or monitoring IT. The objective of the standard is to make sure that the use of IT is effective, efficient and acceptable. The ICT management requires governance function, which has three main tasks in ISO/IEC 38500 [7]:

- "Evaluate current and future use of IT".
- "Direct preparation and implementation of plans and policies to ensure that use of IT meets business needs".
- "Monitor conformance to policies, and performance against the plans"

However, six principles that are mapped into the governance tasks as follows (adapted from ISO/IEC 38500 and SFS [8]):

- All tree tasks are mapped within
  - Responsibility - "Organizations members and groups understand their own responsibilities for IT"
  - Acquisition - "The decision making of IT sourcing is transparent and clear. The correct balance is ensured between benefits, opportunities, cost and risks"
  - Performance - "IT is fit to answer needs of the Business"
- The evaluation and direct tasks are mapped within
  - Strategy - "The business strategy of the organization takes note on the current and future possibilities of the IT. IT strategy offer answers to current and future needs of Business"
  - Human Behaviour - "IT policies, practices and decisions respect current and future human behaviour"
- The evaluation and monitor tasks are mapped within
  - Conformance - "IT complies with legislation and regulations"

2.2   ITIL

ITIL (Information Technology Infrastructure Library) is a widely used framework. It is a collection of the guidelines and best practices involving the ICT services. ITIL recognizes ISO/IEC 38500 as a standard for the governance of ICT. The items that are used to evaluate the organization are financial performance, service and project portfolios, ongoing operations, escalations, opportunities and threats, proposals, contracts and feedback. Directing includes delegation of authority and responsibility; steering committees; vision, strategies and policies are communicated to managers; decisions that have been escalated to management. Monitoring requires a measurement system, key performance indicators, risk assessment, compliance audit and capability analysis. Shortly, governance ensures "that policies and strategy are actually implemented, and that required processes are correctly followed" and it "includes defining roles and responsibilities, measuring and reporting, and taking actions to resolve any issues identified". There are five processes that are related to the ICT governance [9]:

- Strategy management for IT services is used "to review and plan on an annual basis"; "to analyse, set objectives, perspectives, positions, plans and patters for new business or service opportunities". Strategy management for IT services will "assess the impact of environmental changes on the existing strategic and tactical plans", as well as, "articulate what business outcomes need to be met and how services will accomplish this".

- Financial management for IT services "provides the information and history to enable the service provider to determine the value of the investment in a service"; "help implement and enforce policies that ensure the organizations is able to store and archive financial data, secure and control it and make sure that it is reported to appropriate people" (i.e., compliance); "to make sure that investments are appropriate for the level of service that customers demand, and the level of returns that are being projected" (i.e., cost optimization); provides "financial data and information to quantify the potential effect on the business. . . helps to quantify and prioritize the actions that need to be taken to prevent the impact from becoming reality" (i.e., business impact analysis).

- Service portfolio management ensures that "the service provider has the right mix of services to balance the investment in IT with the ability to meet business outcomes".

- Demand management is "the process that seeks to understand, anticipate and influence customer demand for services and the provision of capacity to meet these demands".

- Business relationship management enables to provide links between the service provider and customers. The links ensure "the service provider understands the business requirements of the customer and is able to provide services that meet these needs".

Objectives, triggers (e.g., events that require actions), roles, inputs and outputs are not explicitly related to the phases of the processes. However, it is possible to map triggers, inputs and outputs within the phases of the process.

2.3   COBIT

COBIT gives to the organization tools for forming objectives for the IT function and ways to achieve them. The framework helps organization to get most benefits out of IT while taking note on risks and used resources. COBIT was developed in order to recognize mistakes and to prevent them. The earlier versions of the framework focus on organizations business processes. The COBIT framework (Fig. 2) makes clear separation between governance and management and their processes. There are five domains of the processes for ensuring the governance of the enterprise IT [10]: governance framework setting and maintenance, benefits delivery, risk optimisation, resource optimisation and stakeholder transparency. Each process for governance has its own process goals and the common IT-related goals that relate to multiple governance processes. COBIT describes metrics to be used on goals level, For example, the goal "Individual IT-enable investment contribute optimal value" can be measured with metrics: "Level of stakeholder statisfaction with progress towards identified goals, with value delivery based on surveys" and "Percent of expected value realised".

We found out the governance-related processes by adapting the COBIT enabling processes [11]. At first, we took the process practices, activities, inputs as a base and then collected information about the governance processes. Furthermore, the roles that have responsibilities related to the governance processes are collected. In order to make it easier to see how and where the governance processes communicate and where they get inputs, we collected the processes and practises from their respecting tables.

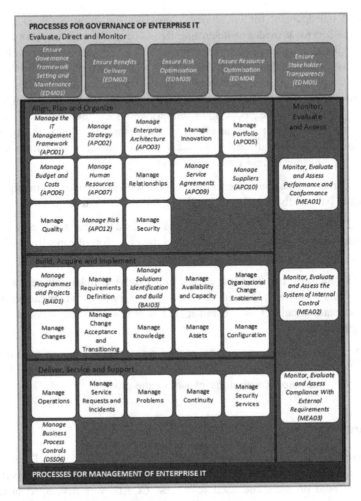

Fig. 2. COBIT in where the processes that have relationships with governance are marked with italics and the corresponding abbreviations are added

2.4 ICT Standard for Management

The first edition of IT Standard or IT Standard for Business [12] was published in 2009 under the title of ICT Standard for Management [13]. In addition guidelines, that take note on specifics of public administration, have been made for the ICT Standard for Management [14]. The ICT Standard for Management divided the ICT functions into five streams: strategy and governance, sourcing and vendor relationships, project management, service management and business alignment, which combines all the other streams together (Fig. 3). Furthermore, we found the definition for the governance modes from the Strategy, Governance and Procedures function as follows: "The governance model defines ICT management related responsibilities, obligations,

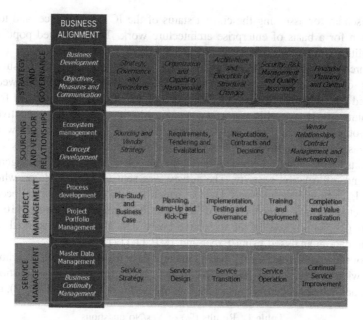

Fig. 3. Streams of the ICT Standard for Management. The functions marked with italics have relationships with the ICT governance.

reporting and communication practices as well as the decision-making model, not only within the ICT function, but also between ICT and its interest groups".

We assumed that the functions of the strategy and governance streams belong to the governance-related functions. Furthermore, we selected the other governance-related functions by related mandatory tools for organization, operative management and development and practices. The Contract portfolio tool was mapped within the Vendor Relationships, Contract Management and Benchmarking function and the System chart tool the definition of which is "Describes all parts of the company's information system architecture as well as intersystem interfaces and data flows" was mapped within the following functions: Sourcing and Vendor Strategy, Business Continuity Management, and Concept Development.

The governance-related functions of the ICT Standard for Management did not map roles for specific functions. The roles were described on the stream level and the framework does not go into details with measurements. Furthermore, the objectives were handled as mention in the Objectives, Measures and Communication function.

3 Survey of the ICT Governance in Northern Savonia

In Finland, the Act of Information Management Governance in Public Administration [15] came into force 2011, and it imposes obligations on how the ICT governance must be designed and described in the public sector. The law focuses on interoperability and mandates the public organizations to make enterprise architecture descriptions and manage them. In Northern Savonia the federation of the municipalities decided to order survey research for assessing the current status of the ICT governance, and to gather information for a basis of enterprise architecture work. The combined population of municipalities is about 50 000 in Northern Savonia.

Research material was gathered by an online survey. A planning of the survey and designing of questions began at January 2014, survey was executed between 18th March and 11th June 2014, and afterwards the research data was analysed from different points of view. In this paper, we separate the answers of the executive board from the others because their role is important in the ICT governance. However, the survey was targeted to those who work on the executive boards, heads of those departments that use the ICT services and those responsible for the IT management, IT security and information security. The survey was sent to 331 people on whom 136 responded. The response rate was 41.1 %, which is a good response rate, because on the internet surveys, a response rate of 30 % can be considered adequate [16]. 28 respondents (21 %) worked on the executive board and 108 (79 %) worked on other roles. The data is analysed using Excel and JMP 11 [17].

One of the goals of the survey was to make it easy for the respondents who are not familiar with the ICT governance. We mainly used forced binary questions [18] (Table 1) and when needed, we also added an option for don't know (Table 2), trying to

Table 1. Results (%) of Yes/No questions

Stream/question	Other (n = 108)		Executive board (n = 28)	
	Yes	No	Yes	No
Business alignment				
Are decision rights defined clearly?	83,33	16,67	85,71	14,29
Is there information about current and future development projects in the municipality?	29,63	70,37	67,86	32,14
Is there successful collaboration in development between business and ICT functions?	37,96	62,04	53,57	46,43
Strategy and governance				
Is there enough information for making decisions and directing daily work?	64,81	35,19	78,57	21,43
Is the Act of Information Management familiar?	30,56	69,44	39,29	60,71
Is the concept of enterprise architecture familiar?	30,56	69,44	57,14	42,86
Is there a need for ICT strategy?	65,74	34,26	60,71	39,29
Is there a need for the ICT governance model?	86,11	13,89	82,14	17,86
Are there tasks that should be done in collaboration with the ICT Function in the municipality?	48,15	51,85	64,29	35,71
Is the resourcing of the ICT function sufficient?	50,00	50,00	46,43	53,57
Sourcing and vendor relationships				
Is the ICT function consulted before development projects?	59,26	40,74	85,71	14,29
Are the current information system contracts familiar?	12,04	87,96	32,14	67,86
Service management				
Are information systems used efficiently?	58,33	41,67	60,71	39,29
Is there a standard operation model for data processing?	37,96	62,04	64,29	35,71

Table 2. Results of Yes/Don't know/No questions

Stream/question	Other (n = 108)			Executive board (n = 28)		
	Yes	Don't know	No	Yes	Don't know	No
Business Alignment						
Is business development successful through IT in the municipality?	42,59	47,22	10,19	57,14	28,57	14,29
Strategy and governance						
Are there objectives for ICT function in the municipality?	46,30	46,30	7,41	67,86	21,43	10,71
Are there objectives for information security in the municipality?	63,89	27,78	8,33	78,57	10,71	10,72
Are information security policies integrated into daily work in the municipality?	65,74	24,07	10,19	75,00	14,29	10,71
Is ICT noted on contingency planning in the municipality?	17,59	75,93	6,48	50,00	46,43	3,57
Are the daily tasks of ICT function taken into consideration in the organization model in the municipality?	48,15	46,30	5,56	67,86	25,00	7,14
Sourcing and Vendor Relationships						
Are there objectives for ICT sourcing in the municipality?	39,81	50,00	10,19	53,57	28,57	17,86
Project management						
Are there objectives for ICT projects in the municipality?	25,93	63,89	10,19	42,86	39,29	17,86
Are projects managed sufficiently in the municipality?	17,59	75,00	7,41	35,71	46,43	17,86
Service management						
Are there objectives for ICT services in the municipality?	30,56	62,04	7,41	53,57	25,00	21,43
Are there objectives for data management in the municipality?	50,93	42,59	6,48	75,00	10,71	14,29

eliminate the respondents need to make guesses. The questions might have an optional text field for expanding the answer. Our mappings between the presented framework-based governance and the established questions [6] are mainly based on the generic features (i.e., objectives and roles) of the presented frameworks within the following reasons:

- We used the streams of the ICT Standard for Management when we grouped our questions. The objectives realize goals and are tracked against measures. Therefore, we asked about stream-based objectives as follows:
  - Are there objectives for the ICT function?
  - Are there objectives for the information security?
  - Are there objectives for the ICT projects?
  - Are there objectives for the ICT services?
- There are several roles in the frameworks. However, instead of the roles we asked the questions the main aims of which were to assess knowledge about responsibilities the options of which are mainly Do co-operation:
  - Who is responsible for the planning and requirement specification during the development of an information system?
  - Who is responsible for the tendering during the development of an information system?
  - Who is responsible for sourcing and contracts during the development of an information system?
  - Who is responsible for deployment during the development of an information system?
  - Who is responsible for continual service improvement after the development of an information system?
  - Who is responsible for the ICT budgeting? The options are Field of operation, ICT function or In co-operation
  - Who coordinates sourcing of information systems? The options are Executive board, ICT function or Someone else
- Collaboration between business and ICT, as well as, the decision making is carried out according to the governance model. Therefore, we asked the following questions:
  - Are decision rights defined clearly?
  - Is business development successful through IT?
  - Is there successful collaboration in development between the business and ICT functions?
  - Are information security policies integrated into daily work?
  - Is there a need for the ICT strategy?
  - Is there a need for the ICT governance model?
  - Is the resourcing of the ICT function sufficient?
  - Is the ICT function consulted before development projects?
- The operational model includes a set of rules and guidelines for operations, it also includes the main idea, main actors and main resources (information, skills, tools) needed in operation [19]. The operational model must also support achieving the objectives and generating the products (i.e., the outputs of the operations). Therefore, we asked the following questions
  - Are the daily tasks of the ICT function taken into consideration in the organization model?
  - Are projects managed sufficiently?

- Are information systems used efficiently?
- Is there a standard operation model for data processing?
- Both governance and operations require information, which is formed from data sources that are either inside or outside of the organization. Therefore, we asked the following questions:
  - Is there information about current and future development projects?
  - Is there enough information for making decisions and directing daily work?
  - Are there objectives for data management?
- There was one specific need, i.e., how ICT is noted on contingency planning. Furthermore, there were need to know if the Act of Information Management is familiar and is the concept of enterprise architecture familiar.

There is clear difference how familiar the respondents are with the current and future development projects. The members of the executive board have better knowledge about the development projects than the respondents in other roles. There were a few questions in which one answer alternative was "Don't Know". The members of the executive board have less uncertainty (i.e. Don't Know answers) on their responses. Furthermore, the response groups differ on the questions related to the ICT objectives. The members of the executive board were familiar with the objectives.

There were five questions the main aims of which were to assess knowledge about responsibilities (Don't know, Field of operation, ICT function, ICT service provider, In co-operation). The members of the executive board have better knowledge about the responsibilities during information system development than those who work on others roles as follows:

- Who is responsible for the planning and requirement specification during the development of an information system?
  - Executive board: 7,14 % Don't know, 39,29 % Field of operation, 7,14 % ICT function, 7,14 % ICT service provider, 39,29 % In co-operation
  - Other: 20,37 % Don't know, 34,26 % Field of operation, 7,14 % ICT function, 4,63 % ICT service provider, 33,33 % In co-operation
- Who is responsible for the tendering during the development of an information system?
  - Executive board: 7,14 % Don't know, 7,14 % Field of operation, 28,57 % ICT function, 28,57 % ICT service provider, 28,57 % In co-operation
  - Other: 26,85 % Don't know, 6,48 % Field of operation, 21,30 % ICT function, 31,48 % ICT service provider, 13,89 % In co-operation
- Who is responsible for sourcing and contracts during the development of an information system?
  - Executive board: 7,14 % Don't know, 7,14 % Field of operation, 25,00 % ICT function, 28,57 % ICT service provider, 32,14 % In co-operation
  - Other: 25,00 % Don't know, 9,26 % Field of operation, 21,30 % ICT function, 22,22 % ICT service provider, 22,22 % In co-operation
- Who is responsible for deployment during the development of an information system?
  - Executive board: 7,14 % Don't know, 14,29 % Field of operation, 10,71 % ICT function, 28,57 % ICT service provider, 39,29 % In co-operation

- Other: 15,74 % Don't know, 13,89 % Field of operation, 15,74 % ICT function, 21,30 % ICT service provider, 33,33 % In co-operation
- Who is responsible for continual service improvement after the development of an information system
  - Executive board: 17,86 % Don't know, 21,43 % Field of operation, 10,71 % ICT function, 21,43 % ICT service provider, 28,57 % In co-operation
  - Other: 27,78 % Don't know, 25,00 % Field of operation, 8,33 % ICT function, 7,41 % ICT service provider, 31,48 % In co-operation

When we asked who (Executive board, ICT function or Someone else) coordinates the sourcing of information system then it can be concluded that the ICT function coordinates the sourcing of information systems (EB 57,14 %, Other 48,1 %). There were three options (Field of operation, ICT function, In co-operation) of the answers to the question who is responsible for ICT budgeting. The ICT budgeting is the mostly done in co-operation both in the municipality (EB 46,43 %, Other 58,33 %) and in the field of operation (EB 29,29 %, Other 63,89 %), The executive board also emphasizes co-operation in the ICT budgeting, but also notices the responsibility of the ICT function both in the municipality (EB 35,71 %, Other 24,07 %) and in the field of operation (EB 37,71 %, Other 12,04 %).

4 Discussion

There are differences between the frameworks. COBIT describes the processes for the ICT governance explicitly. However, some practices do not have inputs and there are no triggers. ITIL lists items for the processes that are related to the ICT governance. The ICT Standard for Management gives several tools that can be used to find out governance-related functions. In generally, the frameworks present the elements of different kind around practices illustrating, for example, principles, processes and roles. Some frameworks, for example ITIL, list the elements (e.g., objectives, triggers, roles, inputs and outputs) and the listed elements are not necessarily related to the processes.

Surveys can be use in change management - at the same time we can figure out capabilities and give some conceptual information around the survey area. It is very important to customize the questions of the survey based on the elements of the selected frameworks. Moreover, it is important to formulate the questions respondent friendly.

The contingency analysis was used to find out statistical dependencies within the responses. If there is not statistical dependence between the answers (Fisher Exact Test probability > 0,05), one should consider whether those should support each other. Regardless of the response group, there is a strong dependency between the objectives for the ICT project and if the projects are managed sufficiently.

When a contingency is analyzed around the Act of Information Management and contingency planning we made some observations. One important objective of the survey was to prepare respondents to forthcoming enterprise architecture work, which is made mandatory by the Act of Information Management Governance in Public Administration. The results show (Table 3) that familiarity with the Act of Information Management and the concept of enterprise architecture is connected and that currently

Table 3. Contingency analysed questions

Stream/contingency analysed question pair		Fisher's Exact test prob. (P)			
		Municipality		Field of operation	
		Other (n = 108)	Executive board (n = 28)	Other (n = 108)	Executive board (n = 28)
Business alignment					
Is there information about current and future development projects?	Is business development successful through IT?	0,0058	0,5337	0,4146	0,2071
Strategy and governance					
Are there objectives for the ICT function?	Is there a need for the ICT strategy?	0,4715	0,1869	0,3437	1,0000
Are there objectives for information security?	Are information security policies integrated into daily work?	0,0185	0,3063	0,0010	0,7651
Are there tasks that should be done in collaboration with the ICT function?	Are the daily tasks of the ICT function taken into consideration in the organization model?	0,0002	0,0012	0,0013	0,0009
Sourcing and vendor relationships					
Are there objectives for the ICT sourcing?	Are the current contracts on the information system familiar?	0,9089	0,6458	0,6380	0,3158
Project management					
Are there objectives for ICT projects?	Are projects managed sufficiently?	<0,0001	0,0031	<0,0001	0,0029
Service management					
Are there objectives for data management?	Is there a standard operation model for data processing?	0,0043	0,3655	0,0082	0,5351
Are there objectives for the ICT services?	Are information systems in efficient use?	0,3878	0,8776	0,0792	0,5315

they are both still unfamiliar, regardless of the role of the respondent (Fisher's exact test 0,0060 for EB and 0,0402 for Other). There is also considerable lack of awareness about the state of the contingency planning of ICT, more than half of the respondents were not familiar about how ICT is noted on the contingency planning on either municipality or in the field of operation. Fisher's exact test gave less than 0,0001 both the executive board and others.

Our survey in Northern Savonia established that the binary questions are respondent-friendly. Moreover, the set of questions illustrates the scope of the governance function and partly the scope of the management function. The answers revealed several improvement areas. The decision making is not carried out according to the governance model. Moreover, the objectives, that realize the goals and are tracked against measures, are either unknown or they do not exist. Because the respondents are unfamiliar with the enterprise architecture, there is no preparedness to make the enterprise architecture descriptions.

References

1. itSMF-NL: Frameworks for IT Management. Van Haren Publishing, Amersfoort (2006)
2. The Open Group: TOGAF® 9.1, Part VII: Architecture Capability Framework, Architecture Governance (2014)
3. ISACA: Cobit 5: A Business Framework for Governance and Management of Enterprise IT. Printed in United States of America (2012)
4. TOGAF® 9.1. Part IV: Architecture Content Framework - Content Metamodel. http://pubs.opengroup.org/architecture/togaf9-doc/arch/chap34.html
5. ISO/IEC 38500:2015. Information technology - Governance of IT for the organization. https://www.iso.org/obp/ui#iso:std:iso-iec:38500:ed-2:v1:en:term:2.9
6. Meriläinen, H., Hotti, V., Isoniemi, E., Juvonen, P.: Tietohallinnon ohjauksen merkityksen organisaatiotasoinen arviointi. Finnish J. eHealth eWelfare (2015)
7. ISO/IEC 38500:2015. Information technology - Governance of IT for the organization
8. SFS: Corporate Governance of Information Technology ISO/IEC 38500:2008. Finnish Standards Association SFS (2014). http://www.sfsedu.fi/files/122/ISO-38500.ppt
9. Cabinet office: ITIL service strategy. Printed in UK (2011)
10. ISACA: Cobit 5: A Business Framework for Governance and Management of Enterprise IT. Printed in United States of America (2012)
11. ISACA: COBIT 5: Enabling processes. Printed in United States of America (2012b)
12. https://www.itforbusiness.org/book/ict-standard-for-management/preface/
13. ICT Standard Forum: Tietohallintomallin soveltamisohje julkiselle hallinnolle, Laine Direct Oy (2013)
14. L 10.6.2011/634. Act of Information Management Governance in Public Administration. Ministery of Justice, Finlex. http://www.finlex.fi/fi/laki/ajantasa/2011/20110634. Available also in english: http://www.vm.fi/vm/en/04_publications_and_documents/03_documents/20110902ActonI/Tietohallintolaki_englanniksi.pdf
15. University of Texas. Assess Teaching, Response rates. Instructional Assessment Resources. The University of Texas at Austin. https://www.utexas.edu/academic/ctl/assessment/iar/teaching/gather/method/survey-Response.php
16. http://www.jmp.com/software/pro/
17. Dolnicar, S., Grün, B., Leisch, F.: Quick, simple and reliable: forced binary survey questions. Int. J. Market Res. **53**(2), 231-252 (2011). http://ro.uow.edu.au/commpapers/782/
18. Dolnicar, S.: Asking Good survey questions. J. Travel Res. **52**, 551 (2013). http://jtr.sagepub.com/content/52/5/551
19. Innokylä: Mikä on Toimintamalli? Innokylä, National Institute for Health and Welfare (2014). https://www.innokyla.fi/kehittaminen/toimintamalli

Healthcare as a Business Environment – Analyzing the Finnish Health Technology Industry

Reetta Raitoharju[1(✉)], Tuomas Ranti[2], Mikko Grönlund[2], and Kaapo Seppälä[2]

[1] Turku University of Applied Sciences, Turku, Finland
`retta.raitoharju@turkuamk.fi`
[2] University of Turku, Turku, Finland
`{tuomas.ranti,mikko.s.gronlund,kamise}@utu.com`

Abstract. Health technology is the fastest growing high technology field in Finland. However, developing new products, getting international or finding new markets can be challenging in the highly regulated and rapidly changing healthcare environment. This paper presents issues in healthcare as a business environment. A financial analysis was conducted to get an overview about the current stage of Finnish health technology industry. A survey was also conducted in order to find out the challenges Finnish health technology companies are facing. The preliminary results of this financial analysis and survey are reported and then discussed.

Keywords: Health technology · Financial analysis · Regulatory environment

1 Introduction

Healthcare can be seen as a very promising field for new technology: the costs of healthcare should decrease, the personnel is highly educated, and the pressure of taking care of the ageing population demands new solutions. The international market is also huge and while other high technology industries in Finland are suffering from economic depression, health technology has been growing for many years in a row and in 2015, exports of Finnish health technology reached a record of 1.92 billion euros [1].

Despite being an interesting domain, the healthcare sector can be seen as a challenging environment for innovations and new technology. Public healthcare sectors often have strict procurement procedures and the decision making takes place on a high level of organization. Moreover, the regulatory environment is very strict, R&D takes a lot of time and money and medical science develops fast.

In this study, we want to explore the landscape of Finnish health technology companies. The purpose is to map what types of companies this field consists of, and what kind of challenges the companies are facing when it comes to e.g. regulatory issues or financing. This paper first draws an overview of the Finnish health technology industry and its companies. Then we introduce healthcare as a business environment by presenting a business model framework and discussing its factors in the light of healthcare

© Springer International Publishing Switzerland 2016
H. Li et al. (Eds.): WIS 2016, CCIS 636, pp. 207–220, 2016.
DOI: 10.1007/978-3-319-44672-1_17

specificities. Then we describe the methods we used in our study. Finally we will present the preliminary results of our financial analysis and survey and discuss our findings.

2 Finnish Health Technology Industry

Health technology is considered as a strongly developing sector that comprises of several different business segments. The term *health technology* covers all the devices, information systems, equipment, diagnostics and services that are used in self-care, self-monitoring, diagnostics and monitoring in the healthcare environment and the means to compensate disability or malfunction. Healthcare information systems are also part of the health technology field with software, consulting and services such as eHealth and mHealth. In this paper, health technology is understood as medical equipment and devices (e.g. x-ray equipment, imaging devices), medical software (e.g. patient health records) and diagnostics and reagents.

Health technology companies are a very diverse group of businesses, varying from large manufacturers of hospital beds to in vitro diagnostics startups. In the trade statistics the health technology industry can be classified in the following four segments and subsegments:

- Medical equipment: electromedical equipment (e.g. patient monitors), accessories and supplies, instruments and appliances, ventilation equipment, x-ray (including dental), and other imaging equipment.
- Medical Furniture: hospital operating tables and beds, dental chairs and furniture, sterilization equipment, rehabilitation equipment.
- Medical Implants: permanently implanted devices, wearable devices.
- In Vitro Diagnostics: instruments, reagents, supplies [1].

2.1 Exports

In Finland, health technology exports are growing strongly and reached a new record level of €1.92 billion in 2015, showing a product export growth of 6.6 % with a trade surplus of €896 million. The biggest contributor to the trade surplus was the medical equipment segment with 89 % of the total. The second biggest contributor was the In Vitro Diagnostics segment. The long term growth in the sector is also significant; a twenty-year analysis shows an average growth of 6.2 % in exports [1].

In 2015 the biggest destination for Finnish health technology export was North America with 39.0 % of the total. Exports to Europe contributed to 35.5 %, Asia 13.5 %, South and Central America 3.0 %, Oceania 2.5 % and Russia 2.4 %. The biggest destination country was USA with 35.6 % of total exports. The second biggest country was Germany with 11.3 % and the third China with 7.7 % [1].

2.2 Regulation

Health technology companies must follow regulatory requirements in order to place their product in the market. In the European Union, companies need a CE marking whereas in other countries and continents the requirements vary. In order to get into a certain market, a manufacturer must follow the regulations. In the European Union, the defining directive is the Medical device directive that defines the outlines of product development. In the USA, one must follow the FDA regulations, in Canada the CMDCAS approval; in Japan the JPAL law, in Australia the TGA regulation etc. Although there has been an attempt to harmonize these authority regulations across countries differences that must be taken into consideration still exists. Technical procedures and processes are highly standardized, for instance IEC 62304 (Medical device software – Software life cycle processes, ISO 14971 (Medical devices – Risk Management) and IEC 62366 (Medical devices – Usability Engineering). The authority approvals of health technology products have been shifting from testing to quality systems and their auditing [2].

2.3 Studies About Finnish Health Technology Industry

Some studies have been made on Finnish health technology companies. A recent study [3, 4] explored Finnish health technology companies and their current stage, trends and future needs and the ways in which the public healthcare and health technology companies could work together. In these studies, health technology companies estimated their strength in business to be service and product delivery, customer management, quality management and R&D. The weakest part was international business, strategy, product life-cycle, digitalization of business knowhow and authority approvals and regulatory issues.

A master's thesis about the long-term success factors of Finnish health technology companies found the innovative and high-quality products to be the key to success. In addition, effective sales and marketing operations as well as networking in the international business field were found to be significant success factors [5].

3 Healthcare as a Business Environment

Healthcare differs from traditional business-to-business or business-to-customer fields in many ways and traditional business models may not be suitable in many cases. The healthcare environment has unique features when it comes to the technological environment, market environment, regulatory environment and investment environment. In this paper, we use the business model framework (Fig. 1) to map the environment in which health technology companies are operating.

3.1 Market Environment

In the health technology environment, the customers are in many cases institutional. For instance hospital beds or diagnostics are most likely sold directly to hospitals, healthcare centers or other such institutions. When it comes to private customers, there is a lack of retail market in the healthcare sector and the consumers (patients) are not shopping for products and services that best meet their needs as in many other fields of business [7]. However, the healthcare environment is undergoing some major changes.

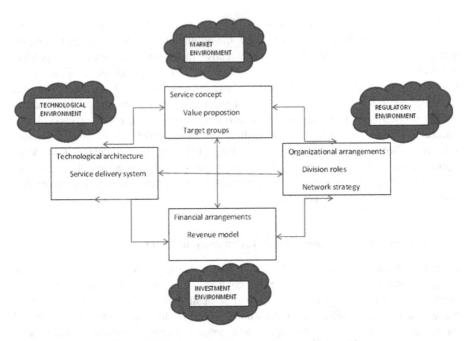

Fig. 1. Business model framework for healthcare [6]

One of the big issues is the consumerization of healthcare. Patients have greater choice in where to go. This leads to the need for healthcare organizations to increase the quality and effectiveness of care. This could offer new business opportunities also for health technology companies.

3.2 Regulatory Environment

The healthcare sector is guided by several regulations that define what can be done. Moreover, regulations prescribe who can do it, where it can be done, when, how and why. The purpose of strict regulatory demands is to protect the patients since health technology has the potential to irreversibly harm the patient. However, there has been a long debate on whether the standards and regulations can keep up with health technology or if they inhibit the development [8, 9]. There are even some critics that argue that health technology innovation can be hurt by the regulatory system [10]. For health technology companies, mastering the regulatory environment can become an advantage. Adapting to the new requirements faster than the competitor will create a competitive edge. Knowhow about the regulatory environment in different countries can help in internationalization and create savings in product development.

3.3 Technological Environment

Medical technology is developing fast and new ways of diagnosis and treatment emerge constantly. This creates new possibilities for the health technology companies that should follow their customers and their changing needs closely. Healthcare is an intolerant sector for mistakes and errors in technology. Also the so called "human error" is considered as a consequence of a failure of one or several parts of the product [11]. Therefore, usability plays an important role in the development of health technology.

3.4 Investment Environment

Health technology development is a risky business that often takes a lot of time (5 to 10 years even) and money. Even then it is estimated that around 50 % of medical device patents never end up in a commercialized product [12]. Therefore R&D in health technology requires investors that have the patience and strong belief in the product. Since this can be a challenge especially in the very early stages in innovation, companies in health technology often need support. Therefore, most industrialized countries actively support the development of health technology and innovations [13].

4 Data and Methods

4.1 Financial Analysis

For the financial data, Bureau van Dijk Electronic Publishing's Orbis Europe database was utilized. The database contains the financial information of over 80 million European companies. Once the companies of the health technology industry were identified, as a second step of the research, a financial statement analysis of the data set was carried out. The period of analysis spans the years 2007–2014.

Financial statement analysis is defined as the process of identifying the financial strengths and weaknesses of the firms by properly establishing a relationship between the items of the balance sheet and the profit and loss account. Financial statement ratio analysis provides a broader basis for comparison than raw numbers do. However, ratios on their own, without year-to-year or other industry or firm-related comparative data are of little use in judging the health or future of the industry or firm being analyzed. There are various methods and techniques that are used in analyzing financial statements, such as comparative statements, schedule of changes in working capital, common size percentages, funds analysis, trend analysis and ratios analysis. In this paper, the objectives of the financial statement analysis were to: provide a picture of the company's profitability and financial status that was as real and accurate as possible and enable comparison of different financial years. In this research, we have measured several dimensions of the financial performance of the companies in the health technology industry in Finland. These dimensions include the extent of operations, the profitability of operations, the solidity of the company, and productivity. Profitability ratios measure the results of business operations or overall performance and effectiveness of the company. Leverage ratios or long term solvency ratios convey a firm's ability to meet the interest costs and payment schedules of its long-term obligations. Productivity ratios give information about the effectiveness and efficiency of personnel and capital use.

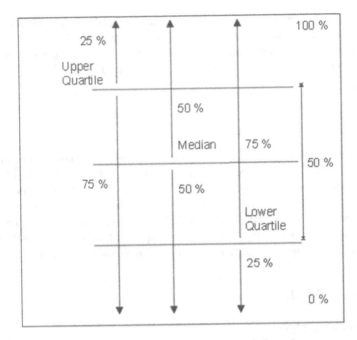

Fig. 2. Statistical figures of location and dispersion

The following key figures are used to describe the development of indicators: median, lower quartile and upper quartile. The interpretation of the figures always depends on the size and quality of the data utilized. The general interpretation of these indicators is presented below (Fig. 2). The median is based on placing the indicators in their order of magnitude. It is the number separating the higher half of the data from the lower half. The median of a finite list of numbers is found by arranging them from lowest to highest. The number in the middle of the list is the median. If there is an even number of observations, the median is usually defined as the mean of the two middle values. Researchers agree that the median is a better indicator of the average value of the indicators. The upper quartile, or the 75th percentile, splits off the highest 25 % of the data from the lowest 75 %. The lower quartile, or the 25th percentile, splits off the lowest 25 % of data from the highest 75 %.

4.2 Survey

To get an overview of Finnish health technology companies and their challenges in the health technology environment, we chose to conduct a survey. With a survey, we can reach a large number of companies and make descriptive and exploratory analysis. The questionnaire consisted of questions that were categorized under the main topics of

- Background information
- Product information

- Quality system
- Market entrance
- Customers
- Sales
- Challenges and difficulties
- Future
- Export and internationalization

The questionnaire was developed and piloted with experts and professionals from health technology companies. The survey was decided to be distributed as an electronic survey using the Webropol service.

In order to reach health technology companies in Finland, we used two different channels. First, a list of companies that have registered a health technology product was ordered from Valvira (National Supervisory Authority for Welfare and Health). Second, a member list from FiHTA (The Finnish Health Technology Association) was used in order to get information about the companies that did not have a CE mark but were in the process towards it or had products that did not require to be registered with Valvira. After that, we analyzed the companies and decided to exclude companies that made products only for private use. These companies were mainly dental laboratories. These were excluded because the nature and volume of this particular business was deemed to differ from the other health technology businesses. After that, we ended up with 474 companies. Email addresses were then collected for the companies. We were able to find the correct email addresses of 350 companies, to which emails containing an individual response link were sent. The contact persons were chosen based on their position in the company management and expected expertise in business and regulatory issues.

5 Results

5.1 Financial Analysis Results

In this paper turnover is used as an indicator of size and development among Finland's health technology industry companies. Turnover enables a comparison of the volume of business activities, but comparability varies, even on the company level, due to the different structures of the companies. In assessing the development of turnover, one must also take into account the effect of inflation.

In Finland, the National Supervisory Authority for Welfare and Health (Valvira) maintains a register of companies with CE-marked medical devices. Looking at the sales growth of the companies that have at least one medical device product with a CE-mark, we can see that it has been moderate (Fig. 3). The median growth percent has revolved around 5–6 %, with the exception of a decline in 2008–2009 and a median growth of just 0.7 % in 2012–2013. In 2014, the median number of employees was 13. (Source: Orbis Europe).

Net profit margin (%), also referred to as the bottom line, net income or net earnings is a measure of the profitability of a venture after accounting for all costs. The cumulative net result of a company should be positive for it to be considered profitable.

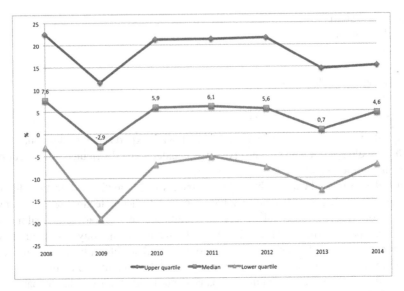

Fig. 3. The growth percent of sales 2008–2014 (group: companies with a CE-marked medical device product, yearly N = 162–219)

A positive net result indicates that the company has been able to cope with interest rates from its regular business operations, as well as managing working capital loans and investment co-financing. The adequacy of net income and the required minimum level is mainly determined by the company's profit distribution objectives. As a rule, the net profit margin is better suited for comparing companies operating in different sectors than EBITDA.

As far as the net result percentages are concerned, we can see that their median has been varying between 1.7 % and 3.8 % in 2007–2014 (Fig. 4). The development in the poorest performing companies in this respect has been on a decline during the past couple of years. The median return on investment has revolved around 7 % during the latest years.

Equity ratio (%) measures a company's capital and reserves/equity, as divided by total assets. The equity ratio measures a company's solidity, ability to tolerate losses and ability to manage long-term liabilities, as the company's asset levels constitute a buffer against any losses. If the capital buffers fall too low, then even one bad year could knock the company down. If a company's profitability is reasonable and stable, but it records high losses, it is considered to have low self-sufficiency. Low equity ratios contain a large risk when profitability is reduced. For this reason, companies should maintain a sufficiently large safety buffer against potential bad years. A high equity ratio also gives a company significantly greater freedom of movement because its dependence on economic cycles and other changes in the operating environment is less pronounced. Companies whose solvency ratios are lower than their competitors' are usually the first to encounter difficulties in a recession. The equity ratio is usually dependent on the age of the company. Young companies are often more indebted than their more established

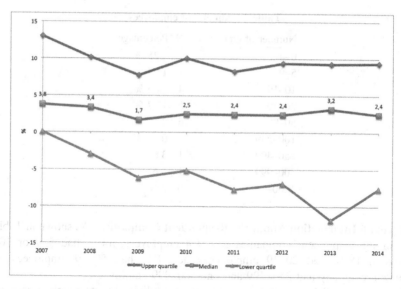

Fig. 4. The development of net result percentage 2007–2014 (group: companies with a CE-marked medical device product, yearly N = 167–236)

counterparts. The indicative norm values for equity ratio (%) are: excellent >50 %, good 35–50 %, satisfactory 25–35 %, poor 15–25 % and weak <15 %.

The median equity ratio of companies with at least one CE-marked medical device product has been good, around 40 % in 2009–2013 and rising to about 45 % in 2014. Measuring the ability of the companies to cover their short-term debt with their liquid assets, the median of their quick ratio has declined somewhat in 2007–2014, but is still at a good level (around 1.2).

Value added refers to the value that is created by the factors of production of a company. Value added is a measure of particular importance, as it has a considerable impact on the ability of an enterprise to pay its workforce and generate profit. The ratio of value added to personnel costs is a way of looking at the relationship between the costs of the people employed in the industry and the value generated. It is a relative measure and therefore not subject to distortions such as exchange rate variations or inflation. Rather than being a direct measure of productivity, it is a measure of the effectiveness of a company in terms of the money spent on employment, regardless of the number of people who are employed. The critical value is 1. If the indicator is below 1, it means the value created does not cover the employment costs, creating an operational loss. Between 2008 and 2013, the median value added per personnel costs of the Finnish health technology companies with a CE-market product declined from 1.36 to 1.27.

5.2   Survey Results

From the first round, we got 29 answers for the questionnaire representing a response rate of 8 %. The preliminary results of this round are reported below.

Table 1. Number of employees

Number of employees	N	Percentage
0–4	7	25 %
5–9	4	14.3 %
10–19	4	14.3 %
20–49	5	17.7 %
50–99	6	21.4 %
100–249	0	0 %
250–499	1	3.6 %
500–999	1	3.6 %
1000–	0	0 %

Background Information About the Respondent Companies. As shown in Table 1, 25 % of the respondent companies had 0–4 employees, 14 % had 5–9 or 10–19 employees, 18 % had 20–49 employees and 21 % had 50–99 employees. One respondent company had 250–499 and one 500–999 employees.

The turnovers of the respondent companies were divided quite evenly across given alternatives (Table 2). Two of the respondent companies did not have any turnover.

Table 2. Turnover

Turnover	N	Percentage
No turnover	2	7.1 %
1–199 999 €	2	7.1 %
200 000–399 999 €	2	7.1 %
400 000–999 999 €	5	17.9 %
1 000 000–1 999 999 €	5	17.9
2 000 000–9 999 999 €	5	17.9 %
10 000 000–19 999 999 €	3	10.1 %
20 000 000–€	4	14.3 %

Technological Environment. Most of the respondent companies had a CE-marked product (75 %, N = 21), Five of the respondent companies did not have a CE-marked product (18 %) and two of the companies estimated to get a CE-mark in 2017 or 2018.

Most of the CE-marked medical products were born as a result of the companies' own research and development actions (83 %, N = 20). Six of the respondent companies had developed a CE-marked product in cooperation with a partner. One of the respondent companies had acquired their CE-marked product through a purchase from another company. One of the respondents did not know how the product was developed.

Market Environment. About one fifth (21 %) of the respondents reported that they are collecting user data (age, gender, contact information etc.) about their products end users and 75 % of the respondents reported that they are not collecting such information. 46 % of the respondents are collecting information about the use of their products from the end users, while 45 % are not.

Most of the products sold among the respondents are for institutional customers (64 %). Only 4 % of the respondents are selling all the products to consumers.

The respondents were asked an open question about how they include customers into their research and development. 19 responses were obtained. The following methods were mentioned:

- Voice of the Customer (VOC)
- Usability testing
- Concept testing
- Pilot use cases
- Ergonomic testing
- Customer feedback
- Partnership

The respondents were asked how clear their company's vision of making money is. There, we got quite a unanimous answer; 75 % of the respondents totally agreed with the statement whiles 25 % partly agreed. The respondents were also asked if they find selling a service as a potential business model for their company's health technology products. Here, the results scattered and 18 % totally disagreed, 29 % partly disagreed, 18 % partly agreed and 32 % totally agreed (Table 3). When asking if selling a product with single payment would be a potential business model for their company most of the respondents partly agreed (36 %) or totally agreed (54 %).

Table 3. Vision of making money

	Totally disagree	Partly disagree	I don't know	Partly agree	Totally agree	Total	Average
Our company has a clear vision of how it plans to make money with its health technology products	0 %	0 %	0 %	25 % (N = 7)	75 % (N = 21)	28	4.75
Selling product as a service is a viable business model for our company in the case of health technology products	18 % (N = 5)	29 % (N = 8)	4 % (N = 1)	18 % (N = 5)	32 % (N = 9)	28	3.18
Selling products through a one-time payment is a viable business model for our company in the case of health technology products	4 % (N = 1)	4 % (N = 1)	4 % (N = 1)	36 % (N = 10)	54 % (N = 15)	28	4.32

Regulatory Environment. Most of the respondents were using quality system ISO 13485 (64 %) or ISO 9001 (50 %). Other quality systems named as being used were OHSAS 18001 and ISO 14001. 18 % of the respondents did not have a quality system. 60 % of the respondents were using the Medical devices – Risk management standard ISO 14971, 33 % were using the Medical device software – Software life cycle processes standard IEC 62304 and 27 % were using the Medical devices – Usability Engineering standard IEC 62366. 40 % of the respondents did not know how to answer to this question.

Respondents were asked if they find some issues in quality systems difficult. It was stated that changing standards require work. For small companies using standards can be very demanding and inflexible. Building a quality system was also seen challenging, as were the different demands in different countries.

Investment Environment. The respondents were asked an open question about what they considered the biggest challenges in exporting their products. Some themes were found in the answers (N = 17). Finding a suitable international distributor was mentioned several times. Also different authority approvals and the registration of the products were seen as a challenge, especially with regard to the time and money this process takes. It was also pointed out in the answers that finding financing in the product development phase is quite easy in Finland, but finding financing for export or marketing is much harder. Especially among small companies, the lack of financial resources is seen as a challenge.

6   Discussion

The preliminary results of the survey showed that even though health technology companies are collecting data about the use of their product, they quite seldom collect data about the actual end-users themselves. This is unsurprising, since there is strict legislation regarding the monitoring and storage of this kind of data. However, possibilities to connect usage data with user data, and share it between applications in a flexible manner, is likely to be of great importance for future advancements in the more effective treatment of various diseases, in accordance with the benefits related to the use of "big data" in general. There may be a need for further amendments in the related legislation in order to secure the privacy of end users, but to still be able to optimize the use of this potential.

About one fifth of the respondent companies did not have a quality system and many issues in quality systems and regulatory environment were seen as challenging. More actions should be taken to help small health technology companies in regulatory issues. Also some financing instruments for internationalization could be developed.

The amount of answers we collected with the survey was quite small and in order to make a more statistical analysis, we need to get more data. Moreover, the field of health technology companies is very heterogeneous, as it comprises very different businesses and various types of companies. Therefore, some clustering should be made in order to get a more in-depth analysis of the Finnish health technology companies.

In terms of the financial performance, the health technology companies with registered products displayed a rather steady sales growth on the whole. With regard to their profitability, it has been steady in the better performing half of the group of companies, but in the lower quartile, the development of profitability has been on the decline. This might be an indication of the polarization of the industry.

There are limitations to a study of this kind. The analysis employed in this article is quantitative and based primarily on key financial figures deriving from the financial statements of the firms. Therefore, it answers to questions related to the "what", "where" and "how many" of high-impact firms within the health technology industry. While an analysis of this kind is useful, it is by no means exhaustive. Knowing where high-impact firms are located, how many exist, and the degree to which they contribute to job creation is helpful to many audiences, including policymakers, industry leaders, academicians and researchers, media organizations, and even high-impact firms themselves.

Acknowledgements. This research has been conducted as part of the TELI project financed by the INKA (Innovative Cities) program and the European Regional Development Fund.

References

1. Finland's health technology trade (2015). http://fihtanews.net/images/stories/industry/2015_eng_fihta_finland_health_tech_trade.pdf
2. Stählberg, T.: Terveydenhuollon laitteiden lakisääteiset määräykset kansainvälisillä markkinoilla – Suomi ja EU fokuksessa, Tekes, Helsinki (2015). https://www.tekes.fi/globalassets/julkaisut/terveydenhuollon_laitteiden_lakisaateiset_maaraykset_opas.pdf
3. Hämäläinen, S., Edlund, L.: Kartoitus terveys- ja hyvinvointiteknologia-alan yritysten liiketoimintaosaamisen tulevaisuuden kehitystarpiesta, Aalto EE (2015). http://www.aaltopro.fi/sites/default/files/terveysteknologia_kartoituksen_tulokset_1.pdf
4. Hämäläinen, S., Edlund, L.: Terveys- ja hyvinvointiteknologiayritysten liiketoimintaosaamisen tulevaisuuden kehitystarpeet ja toimintamalli julkisen terveydenhuollon ja terveys-/hyvinvointiteknologiayritysten yhteistyölle. Selvitysraportti. Aalto EE (2015). http://www.aaltoee.fi/sites/aaltoeefi/files/terveysteknologia_selvitysraportti_1.pdf
5. Vilkman, L.: Suomalaisen terveysteknologian merkittävien innovaatioyritysten pitkän aikavälin menestystekijät. Maisterin tutkinto. Organisaatiot ja johtaminen, Aalto yliopisto, Kauppakorkeakoulu (2012)
6. Lehoux, P., Daudelin, B., Williams-Jones, B., Denis, J.L., Longoe, C.: How do business model and health technology design influence each other? Insights from a longitudinal case study of three academic spin-offs. Res. Policy **43**, 1025–1038 (2014)
7. Hwang, J., Christensen, C.M.: Disruptive innovation in health care delivery: a framework for business-model innovation. Health Aff. **27**(5), 1329–1335 (2008)
8. Vincent, C.J., Niezen, G., Stawarz, K.: Can standards and regulations keep up with health technology? JMIR mHealth uHealth **3**(2), e64 (2015)
9. Contez, N.G., Cohen, G., Kesselheim, A.S.: FDA regulation of mobile health technologies. N. Engl. J. Med. **371**, 372–379 (2014)
10. Longhurst, C.A., Landa, H.M.: Health information technology and patient safety. BMJ **322**, 1096 (2012)

11. Vincent, C.J., Blandford, A.: Maintaining the standard: challenges in adopting best practise when designing medical devices and systems. American Medical Informatics Association (2011). http://www.chi-med.ac.uk/publicdocs/WP037.pdf
12. Mattes, E., Stacey, M.C., Marinova, D.: Predicting commercial success for Australian medical inventions patented in the United States: a cross-sectional survey of Australian inventors. Med. J. Aust. **184**, 33–38 (2006)
13. Niosi, J.: Success factors in Canadian academic spin-offs. J. Technol. Transf. **31**, 451–457 (2006)

Specific Topics of e-Health

Exploitation and Exploration Underpin Business and Insights Underpin Business Analytics

Virpi Hotti[✉] and Ulla Gain

School of Computing, University of Eastern Finland, Kuopio, Finland
{Virpi.Hotti,gain}@uef.fi

Abstract. The revolutionary development in the cognitive computing is connected with natural language processing. For example, the IBM Watson Personality Insights service is research-based and IBM's intuition is that writing always reflects the author's personality. The IBM Watson Personality Insights service provides a list of the behaviors that the personality is likely (e.g., treat yourself) or unlikely (e.g., put health at risk) to manifest. However, the usefulness of the objective insights has to figure out in the business context in where the organizations have to perform and conform they duties. Furthermore, the organizations have to predict the future outcomes within several business analytics. In this paper, the ideas around organizational ambidexterity (i.e., exploitation and exploration) are used to clarify the meaning of the objective insights. The objective insights increase the behavior-centric value propositions, as well as, decrease the number of stakeholder-centric business analytics.

Keywords: Exploitation · Exploration · Insights · Principles

1 Introduction

The insight-driven organizations [1] are increasing in future, as well as, insights as services. The insights are generated for further usage, for example, to cover unknown desires or needs (i.e., the human being does not know yet that his desires and needs are formed by this experiences and they have left their marks on his linguistic expressions). When we have insights from the human being, then we will offer him the first of all experiences within products and services the suitability of which corresponds within the needs and desires (or values) of the human being. When we want to be human- or behavior-centric, then we have to learn to question by the mouth of the human being as follows [2]:

- Advise me (i.e., bring expertise to interactions)
- Alert me (i.e., personalize communication within real-time predictive analytics)
- Ask me (i.e., consult on products, services, and social issues)
- Compare me (i.e., offer peer analytics on virtual channel)
- Educate me (i.e., offer digital online and give tips "in the moment")
- Excite me (i.e., offer unexpected services at unexpected moments)

© Springer International Publishing Switzerland 2016
H. Li et al. (Eds.): WIS 2016, CCIS 636, pp. 223–237, 2016.
DOI: 10.1007/978-3-319-44672-1_18

- Find me (i.e., use visualization and analytics to discover segments)
- Grow with me (i.e., connect data and insights the lives and households)
- Know me (i.e., offer new products and services based on understanding desires and needs)
- Let me choose (i.e., offer optional versus prerequisites, roadmaps versus checkboxes)
- Protect me (i.e., offer multifactor security)
- Trade with me (i.e., give in return better products and services based on sharing data, location, and new ideas)

In the dataism era (or data-ism [3]), the world is controlled by the conceptualization in where the concept is defined "abstract entity for determining category membership" [4]. For example, the personality is under conceptualization by cognitive computing. The IBM Watson Personality Insights service [5] applies "linguistic analytics and personality theory to infer traits" [6] from text (e.g., from social media, enterprise data, or other digital communications [7]) - "IBM's intuition is that writing always reflects the author's personality, regardless of the subject matter" [8]. The IBM Watson Personality Insights service uses a corpus of words that reflect the high or low values of particular characteristics [9] at four levels of strength [10]: weak (100 ~ 1500 words), decent (1500 ~ 3500 words), strong (3500 ~ 6000 words) and very strong (6000 + words). The IBM Watson Personality Insights service enables the objective insights of the human beings. The utilizations of the personality insights may differ from self-study to personalized services such as product recommendations [10], matching individuals such as doctor-patient matching because patients prefer doctors who are similar to themselves, monitoring and predicting mental health such as predicting postpartum and other forms of depression from social media, monitoring radical and rogue elements via social media [11]. There are some applications (e.g., Celebrity Match [12] and Investment Advisor [13]) in where the IBM Watson Personality Insights are meaningful part.

One way to bring the insights closer to the experiments in business is to use the same concepts as in business. Nowadays, the evidence-based decision making is required. ISO 9000:2015 has the principle for the evidence-based decision making and the statement of the principle is the follows [14]: "Decisions based on the analysis and evaluation of data and information are more likely to produce desired results". In this paper, the analysis of data and information refers to the business analytics the meaning of which is to predict the outcomes. The evaluation of data and information refers to the business intelligence the meaning of which is to judge the performance. Furthermore, within the business analytics and business intelligence the ideas around organizational ambidexterity (i.e., exploitation and exploration) have been adapted both to clarify the meaning of the business analytics and the objective insights.

The "original meaning of ambidexterity was an individual's capacity to be equally skillful with both hands" [15]. In 1976, Duncan defines organizational ambidexterity to be the ability of an organization to balance short- and long-term objectives. At beginning of the 90's, March replaces short- and long-term objectives within exploitation and exploration. Bøe-Lillegraven crystallizes previous ambidexterity studies as follows [16]: "exploration is linked to growth whereas exploitation is linked to profits". There are a lot of articles written about organizational ambidexterity. However, there was only one article [16] in where ambidexterity is combined within analytics (Google Scholar, intitle:ambidexterity + intitle:analytics) and it does not handle either business analytics or insights based on cognitive computing. In this paper, the adaption of the organizational ambidexterity is novelty - ambidexterity is combined within business analytics and the objective insights such as the personality ones.

The main aim of this paper is to encourage for experiments around the behavior-centric value proposition based on the objective insights. Therefore, we clarify our ambidexterity adaption within the data- and principle-based extractions and the building blocks of the business models, as well as, we mapped some business analytics within exploitation and exploration (Sect. 2). Furthermore, we clarity whether there are explicit explanations for the personality traits of the IBM Watson Personality Insights service and how we can use them for the planning value proposition (Sect. 3).

2 Ambidexterity and Business

Nowadays data is ennobled for the insights that can be equated within the principles. If we understand the patterns in the data, then it is possible to understand principles [17], and vice versa. There have to be two-way transparency between selected data (i.e., capta "which is taken in analysis" [18]) and principles (Fig. 1). Data have to process from raw data to principles and the principles have to have identified mechanisms (i.e., metrics the sources of which are datasets) that will be used to measure whether the principle has been met or not. By setting metrics for monitor performance, organizations can capture timely information to help drive organizational performance.

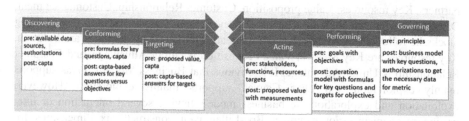

Fig. 1. Data-based and principle-based extractions

When the strategic goals are clear, this means measurable objectives the target of which are set. TOGAF [19] defines the objectives, as time-bounded milestones for enterprises used to demonstrate progress towards goals tracked against measures (i.e., indicators or factors). In generally, the governing body (the "person or group of people who are accountable for the performance and conformance of the organization" [20]) has to understand in where the organizations have to perform and conform they duties, as well as, the governing body familiar with the predictable outcomes within several business analytics. Hence, for example, Bernard Marr has published the world-wide known publications to clarify key questions for analytics and performance [21], key performance indicators [22], key business analytics [23], and even the construction for strategy board having six panels in where the panel-specific questions are as follows [24]:

- The Purpose Panel - why your business exists, and what you want your business to be in the future?
- The Customer Panel - how much you know about the customers, and what you may need to find out in order to deliver on your strategic objective?
- The Finance Panel - how does your strategy generate money, and are you confident your business model is accurate?
- The Operations Panel - what you need to do internally to deliver your strategy, and what core competencies will you need to excel if you are going to execute your chosen strategy?
- The Resource Panel - what resources you need to deliver your strategy and what you may need to find out?
- The Competition and Risk Panel – what is threatening your success?

Instead of panes or canvases the ideas around organizational ambidexterity are used in this paper both clarifying insight-based business and encouraging for experiments around, for example, new technology. Organizational ambidexterity refers to compete both in mature and new markets and technologies – the mature one is labelled with efficiency, control and incremental improvement; the new one is labelled with flexibility, autonomy, and experimentation [25]. Furthermore, there are some alignments (e.g., strategic intent, critical tasks, competences, structure, controls and rewards, culture and leadership role [26]) that have been taken into consideration when exploitation and exploration have been compared.

Business models are described either within building blocks of Business Model Canvas (BMC) or Lean Canvas. The BMC building blocks as follows [27]: Key partners, Key resources, Value proposition, Customer Relationship, Customer segment, Distribution channel, Revenue stream. However, Lean Canvas replaces five building blocks he following way [28]: Key Partners are Problem, Key activities are Solution, Key resources are Key Metrics, Value Proposition is Unfair Advantage, and Customer Relationship is Unfair Advantage. It is obvious that Business Model Canvas supports mainly exploitation and Lean Canvas supports mainly exploration. However, exploitation and exploration are business models having several interaction or integration mechanisms. For example, Bøe-Lillegraven constructs six dimensions for explore and exploit value chains in where activities can be partly the same. Bøe-Lillegraven illustrates the interaction mechanisms between the following six dimensions: resource allocation, cost structure, value proposition, market performance, revenues, and profits. The resource allocation is the first dimension of the value chains and it seems to be challenges for leaders because they have to be "able to orchestrate the allocation of resources between the old and new business domains" [25].

Nowadays, alternative calculations are made around available resources, cost structure, revenues, and profits. Therefore, we do not take optimization things into the consideration, when we want to find out in where the insights based on cognitive computing reduce the number of needed business analytics. When we map different business analytics within exploitation and exploration we used two dimensions (i.e., value proposition and market performance) of the multi-dimensional conceptual

Table 1. Mapping a set of business analytics

Business analytics	Reason for business analytics	Exploitation	Exploration	Value proposition	Market performance
Customer profitability	Finding money making customers	x		x	x
Product profitability	Finding money making products	x		x	x
Value driver	Clarifying value drivers		x	x	
Non-customer	Finding new opportunities		x	x	
Customer engagement	Estimating impacts on the customer experience	x		x	x
Customer segmentation	Increasing revenue by meeting needs		x	x	
Customer acquisition	Finding problems in the buying process		x	x	
Marketing channel	Where and when prospects and customers are reachable?		x	x	

framework of explore and exploit value chains by Bøe-Lillegraven. Therefore, the selected business analytics [23] concern mainly customers and market (Table 1).

Furthermore, if we found the corresponding performance indicators, then we mapped those within market performance. When we made our mappings, we realized that the most of the business analytics concerns the different stakeholders (e.g., customers, employees and shareholders). Therefore, we exemplified the questions in where the personality insights are in centric (Table 2).

Table 2. Mapping personality insights for business analytics

Business analytic	Trait-based question started by "What are the personality insights"
Customer profitability	... of the founded money making customers?
Product profitability	... of the buyers of the founded money making products?
Value driver	... of the most important stakeholders?
Unmet need, Customer segmentation, Customer engagement, Customer acquisition	... of the customers?
Non-customer	... of the prospects?
Marketing channel	... of the customers per marketing channel?

3 Personality Insights

The IBM Watson Personality Insights service provides [10] a personality scoreboard in where the traits are grouped into the personality insights of three kind (i.e., personality, needs and values) and "each trait value is a percentile score which compares the individual to a broader population". The service provides a list of the behaviors that the personality is likely (e.g., treat yourself, click on an ad, follow on social media) or unlikely (put health at risk, re-share on social media, take financial risks) to manifest. The behaviors are based on studies "which reveal different correlations between personality traits and behaviors in certain industries or domains' such as follows [10]: spending habits are related to the emotional range, risk profiles to extraversion, and healthy decisions are related to conscientiousness.

"IBM tends to believe that personality evolves within certain bounds, it has conducted no studies to examine the upper and lower bounds of this evolution" [10]. However, IBM gives the following recommendations [10]: work with the latest data and with as much available data as possible, refresh personality portraits at regular intervals to capture individuals' evolving personalities. Furthermore, the Personality Insights service is evolving. For example, we tested within the IBM Watson Personality Insights service [6] the same text (Fig. 2) twice.

There is the research literature [29] behind the words and they have been validated by testing with users. The get personality traits are divided into three groups as follows [30]:

When we will categorize contents the triggers are formed using operators. The operators are used when the sentiments (i.e., positive, negative or neutral) are recognized, for example, from tweets. Furthermore, it is possible to compare sentences to have an algorithm scoring their similarity.

Nowadays, analytics tools support citizen data scientist, for example, hiding the names of techniques (e.g., algorithms) and instead of statistical issues the business-related concepts are used. For example, data contain data items the role of which (e.g., input or target) can vary. However, data modeling level or type (e.g., nominal, ordinal and continuous/interval) affects for patterns and models. For example, nominal data (e.g., names) can be counted, ordinal data (e.g., ratings) can be counted and ordered, and continuous data (e.g., amounts) can be counted, ordered and measured.

When we explore data then we try to find out patterns or models. The exploring is used of instead of the analytics which refers "to extracting useful business patterns or mathematical decision models from preprocessed data set". If there is no real target to steer the analyzing process, then the main aim is to describe patterns (e.g., associations or clusters). If there is the real target, then the main aim is to build models (e.g., networks, regressions and trees).

Fig. 2. Example text for the IBM Watson personality insights service

- Big Five describe "how a person engages with the world". The model includes the following five primary dimensions (Tables 3, 4, 5, 6 and 7) the six facets of which further characterize an individual according to the dimension.
- Values describe motivating factors that influence a decision making. The model includes five dimensions of human values as follows: Self-transcendence/Helping others, Conservation/Tradition, Hedonism/Taking pleasure in life, Self-enhancement/Achieving success, Open to change/Excitement.
- Needs describe which aspects of a product will resonate with an individual. The model includes twelve characteristic needs as follows: Excitement, Harmony, Curiosity, Ideal, Closeness, Self-expression, Liberty, Love, Practicality, Stability, Challenge, and Structure.

We got partly different summaries (Fig. 3) and there were only two facets, Intellect and Authority-challenging, have decreased one percent from 100 to 99 (Fig. 4). There are explicitly explanations for the Big Five sentences [31]. However, we did not find out explanations for the sentences of Needs and Values. Furthermore, there are a lot of properties [32] for primary and secondary dimensions without explanations for the sentences (e.g., "You are shrewd and inner-directed") based on those properties of the summaries.

There are some traits which characterize that the personality can predict some outcomes. For example, conscientiousness predicts job performance and extraversion

Table 3. Facets of openness (openness is "the extent to which an individual is open to experiencing a variety of activities")

Facet	High explanation	Low explanation
Adventurousness	Adventurous: "You are eager to experience new things"	Consistent: "You enjoy familiar routines and prefer not to deviate from them"
Artistic interests	Appreciative of art: "You enjoy beauty and seek out creative experiences"	Unconcerned with art: "You are less concerned with artistic or creative activities than most people"
Emotionality	Emotionally aware: "You are aware of your feelings and how to express them"	Dispassionate: "You do not frequently think about or openly express your emotions"
Imagination	Imaginative: "You have a wild imagination"	Down-to-earth: "You prefer facts over fantasy"
Intellect	Philosophical: "You are open to and intrigued by new ideas and love to explore them"	Concrete: "You prefer dealing with the world as it is, rarely considering abstract ideas"
Authority-challenging (i.e., liberalism)	Authority-challenging: "You prefer to challenge authority and traditional values to effect change"	Respectful of authority: "You prefer following with tradition to maintain a sense of stability"

Table 4. Facets of conscientiousness (conscientiousness is a "tendency to act in an organized or thoughtful way")

Facet	High explanation	Low explanation
Achievement-striving	Driven: "You set high goals for yourself and work hard to achieve them"	Consistent: "You enjoy familiar routines and prefer not to deviate from them"
Cautiousness	Deliberate: "You carefully think through decisions before making them"	Bold: "You would rather take action immediately than spend time deliberating making a decision"
Dutifulness	Dutiful: "You take rules and obligations seriously, even when they are inconvenient"	Carefree: "You do what you want, disregarding rules and obligations"
Orderliness	Organized: "You feel a strong need for structure in your life"	Unstructured: "You do not make a lot of time for organization in your daily life"
Self-discipline	Persistent: "You can tackle and stick with tough tasks"	Intermittent: "You have a hard time sticking with difficult tasks for a long period of time"
Self-efficacy	Self-assured: "You feel you have the ability to succeed in the tasks you set out to do"	Self-doubting: "You frequently doubt your ability to achieve your goals"

Table 5. Facets of extraversion (extraversion is "a tendency to seek stimulation in the company of others")

Facet	High explanation	Low explanation
Activity level	Energetic: "You enjoy a fast-paced, busy schedule with many activities"	Laid-back: "You appreciate a relaxed pace in life"
Assertiveness	Assertive: "You tend to speak up and take charge of situations, and you are comfortable leading groups"	Demure: "You prefer to listen than to talk, especially in group situations"
Cheerfulness	Cheerful: "You are a joyful person and share that joy with the world"	Solemn: "You are generally serious and do not joke much"
Excitement-seeking	Excitement-seeking: "You are excited by taking risks and feel bored without lots of action going on"	Calm-seeking: "You prefer activities that are quiet, calm, and safe"
Outgoing (i.e., friendliness)	Outgoing: "You make friends easily and feel comfortable around other people"	Reserved: "You are a private person and do not let many people in"
Gregariousness	Sociable: "You enjoy being in the company of others"	Independent: "You have a strong desire to have time to yourself"

Table 6. Facets of agreeableness (agreeableness is a "tendency to be compassionate and cooperative toward others")

Facet	Explanation of summary for high value	Explanation of summary for low value
Altruism	Altruistic: "You feel fulfilled when helping others and will go out of your way to do so"	Self-focused: "You are more concerned with taking care of yourself than taking time for others"
Cooperation	Accommodating: "You are easy to please and try to avoid confrontation"	Contrary: "You do not shy away from contradicting others"
Modesty	Modest: "You are uncomfortable being the center of attention"	Proud: "You hold yourself in high regard and are satisfied with who you are"
Uncompromising (i.e., morality)	Uncompromising: "You think it is wrong to take advantage of others to get ahead"	Compromising: "You are comfortable using every trick in the book to get what you want"
Sympathy	Empathetic: "You feel what others feel and are compassionate toward them"	Hard-hearted: "You think people should generally rely more on themselves than on others"
Trust	Trusting of others: "You believe the best in others and trust people easily"	Cautious of others: "You are wary of other people's intentions and do not trust easily"

Table 7. Facets of emotional range (emotional range is "the extent to which emotions are sensitive to the environment")

Facet	High explanation	Low explanation
Fiery (i.e., anger)	Fiery: "You have a fiery temper, especially when things do not go your way"	Mild-tempered: "It takes a lot to get you angry"
Prone to worry (i.e., anxiety)	Prone to worry: "You tend to worry about things that might happen"	Self-assured: "You tend to feel calm and self-assured"
Melancholy (i.e., depression)	Melancholy: "You think quite often about the things you are unhappy about"	Content: "You are generally comfortable with yourself as you are"
Impulsiveness (i.e., immoderation)	Hedonistic: "You feel your desires strongly and are easily tempted by them"	Self-controlled: "You have control over your desires, which are not particularly intense"
Self-consciousness	Self-conscious: "You are sensitive about what others might be thinking of you"	Confident: "You are hard to embarrass and are self-confident most of the time"
Susceptible to stress (i.e., vulnerability)	Susceptible to stress: "You are easily overwhelmed in stressful situations"	Calm under pressure: "You handle unexpected events calmly and effectively"

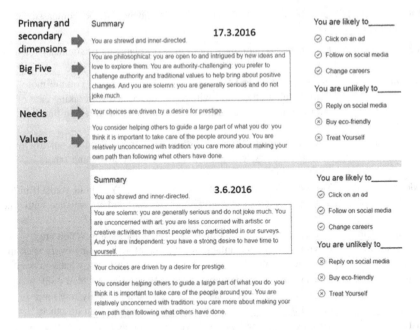

Fig. 3. Examples of summaries

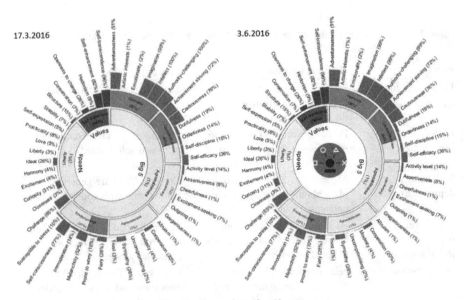

Fig. 4. Examples of sunbursts

Table 8. Examples of mappings between primary characteristics and outcomes

Openness	Conscien-tiousness	Emotional range	Agreeableness	Extraversion	Outcome
High					Try new thing
High					Respond to product respond
	High				Not abuse credits cards
		Low			Abuse credits cards
	Low		Low		Avoid to take risks
				High	Take risks
High	High				Self-improvement
High	High			High	Greater life expectancy
		High			Consume high-fat food
	High				Consume low-fat food
High					Try varied diet
			High		Participate religious practices

indicates job satisfaction. However, we collected 12 outcomes [30] (Table 8) which can be explicitly relate within primary characteristics. Actually, the primary characteristics are called social propensities in the Tone Analyzer service [33].

The Celebrity Match service in where the personality traits (Table 9) are adapted in the way that might decreases the credibility of the IBM Watson Personality Insights service. However, the matching idea comes clear, for example, when the prime minister of Finland has nearly same values as Dalai Lama (Fig. 5).

When we offer experiences within products and services, then it is crucial to understand that the consciousness of the human being is the wholeness of the experiences [34] the contents of which can be qualitatively (consciousness degree, clarity and linguistic to be indicated) different. The human being gets both intentional and unintentional experiences [35]. The intentional experience can be either manifest or latent. The manifest experience will be immediately understood. However, the latent one can be evolved the manifest one, for example, with the help of teaching, growing, psychotherapy or self-assessment. The human being is allowed and he should have unintentional experiences, such as concerts, only for affecting his well-being state. When we are going to arrange or offer experiences of different kind, for example, to our customers, then we adapt the names of the expected experiences as follows: manifest experiences are insightful ones, latent experiences are challenging ones, and unintentional experiences are sensuous (even attractive) ones. When we map the traits (Table 10) within the expected experiences of different kind then we will be behavior-centric when we plan value propositions.

Table 9. Example of adapted personality traits

Extremes of Celebrity match		Personality	Needs	Values
Cautious	Curious	X		
Easy Going	Organized	X		
Reserved	Outgoing	X		
Analytical	Compassionate	X		
Confident	Sensitive	X		
Ease	Challenge		X	
Independence	Belonging		X	
Fulfillment	Exploration		X	
Calm	Excitement		X	
Contentedness	Acceptance		X	
Imperfection	Perfection		X	
Restraint	Freedom		X	
Introversion	Extroversion		X	
Complacency	Eagerness		X	
Obliviousness	Identification		X	
Risk	Stability		X	
Flexibility	Structured		X	
Modernity	Tradition			X
Constancy	Stimulation			X
Stoicism	Hedonism			X
Non-conformity	Conventionality			X
Egoism	Selflessness			X

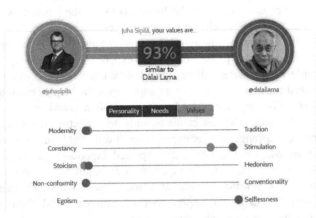

Fig. 5. Celebrity match between Juha Sipilä and Dalai Lama

Table 10. Some personality traits mapped within experiences

Personality trait	Explanation	Insightful	Challenging	Sensuous
Adventurousness	Adventurous: "You are eager to experience new things"	x	x	x
Impulsiveness	Hedonistic: "You feel your desires strongly and are easily tempted by them"			x
Artistic interests	Appreciative of art: "You enjoy beauty and seek out creative experiences"			x
Imagination	Down-to-earth: "You prefer facts over fantasy"	x		
Liberalism	Authority-challenging: "You prefer to challenge authority and traditional values to effect change"		x	
Intellect	Philosophical: "You are open to and intrigued by new ideas and love to explore them"	x		

4 Discussion

We found the zoo of performance indicators and even business analytics. Therefore, we illustrated the business analytics mapping within exploitation and exploration. Furthermore, we took two building blocks (i.e., value proposition and market performance) to fulfill our construction in where we mapped the set of business analytics. When we made our mappings between the business analytics and insights, we realized that the most of the business analytics concerns the different stakeholders (e.g., customers, employees and shareholders). Usually, when the organizations are interested in the engagements of the stakeholders, then the personality insights are reasonable form to them.

When the organization find out balance between exploitation and exploration, it can allocate resources optimally. Hence, within data-driven optimization the understanding organizational ambidexterity, i.e., the organization has to improve and invent at the same time, are needed. The pressures for performance and conformance together with prediction are growing when the organizations either compete or try to be more effectiveness. Therefore, we believe that the understanding the possibilities of the objective insights is crucial. However, the insights have to be actionable ones. Therefore, we constructed both business analytics based mappings within the insights, as well as, the personality traits mappings within the expected experiences (i.e., intentional and unintentional ones the names of which can be modified in value proposition).

If the insights services are going to be used for value propositions, then the insights have to be transparent, i.e., the explanations have to be explicit. Furthermore, the organizations have to form their own points of view for the adaptions of the insights. In future, both experiments and research are needed around the presented constructions.

Especially, the traits of the IBM Watson Personality Insights service are difficult to understand without explanations the reason of why we collected the explanations of Big Five to this paper. The value proposition based on the personality insights might get the new points of view as follows: if you are going to give even provocative experiences then the authority-challenging personalities might be the best audience, if your audience is philosophical ones then well-argument things might be reasonable to offer with a straightforward style. However, the organizations have to make their own experiments if they want to achieve either competitive advantages or effectiveness by adapting the objective insights.

References

1. http://www.kdnuggets.com/2016/01/businesses-need-one-million-data-scientists-2018.html
2. Skibsted, L.A.: Cognitive Business (2016)
3. Lohr, S.: Data-Ism – Inside the Big Data Revolution. Harper Business, New York (2015)
4. ISO/IEC 2382:2015. Information Technology — Vocabulary. https://www.iso.org/obp/ui/#iso:std:iso-iec:2382:ed-1:v1:en:term:2122971
5. IBM.   https://www.ibm.com/smarterplanet/us/en/ibmwatson/developercloud/personality-insights.html
6. IBM. https://personality-insights-livedemo.mybluemix.net/
7. IBM. https://www.ibm.com/smarterplanet/us/en/ibmwatson/developercloud/doc/personality-insights/
8. IBM. https://www.ibm.com/smarterplanet/us/en/ibmwatson/developercloud/doc/personality-insights/guidance.shtml
9. IBM. https://www.ibm.com/smarterplanet/us/en/ibmwatson/developercloud/doc/personality-insights/sample.shtml
10. IBM.   https://developer.ibm.com/watson/blog/2016/01/12/personality-meets-behavior-the-new-ibm-watson-personality-insights-demo/Batista
11. IBM.   http://www.ibm.com/smarterplanet/us/en/ibmwatson/developercloud/doc/personality-insights/usecases.shtml
12. IBM. https://your-celebrity-match.mybluemix.net/
13. IBM. http://investment-advisor.mybluemix.net/
14. ISO 9000:2015 Quality Management Systems — Fundamentals and Vocabulary
15. Birkinshaw, J., Gupta, K.: Clarifying the distinctive contribution of ambidexterity to the field of organization studies. Acad. Manag. Perspect. 27(4), 287–298 (2013)
16. Bøe-Lillegraven, T.: Untangling the ambidexterity dilemma through big data analytics. J. Organ. Des. 3(3), 27–37 (2014)
17. Ahlemeyer-Stubbe, A., Coleman, S.: A Practical Guide to Data Mining for Business and Industry, pp. 35–36. Wiley, Hoboken (2014)
18. Lanigan, R.L.: Capta versus data: method and evidence in communicology. Human Stud. 17(1), 109–130 (1994). Phenomenology in Communication Research
19. TOGAF. http://pubs.opengroup.org/architecture/togaf9-doc/arch/chap34.html
20. ISO/IEC 38500:2015. Information Technology — Governance of IT for the Organization. https://www.iso.org/obp/ui#iso:std:iso-iec:38500:ed-2:v1:en:term:2.9
21. Marr, B.: The Intelligent Company: Five Steps to Success with Evidence-Based Management. Wiley, Hoboken (2010)

22. Marr, B.: Key Performance Indicators: The 75 Measures Every Manager Needs to Know. Wiley, Hoboken (2012)
23. Marr, B.: Key Business Analytics – The 60 + Business Analytics Tools Every Managers Need to Know. Wiley, Hoboken (2016)
24. Marr, B.: From Big Data to Real Business Value: The Smart Approach and 3 More Use-Cases. Wiley (2016). http://www.cgma.org/Magazine/Features/Documents/SMART-strategy-board.pdf
25. O'Really, C., Tushman, M.: Organizational ambidexterity: past, present, and future. Acad. Manag. Perspect. 27(4), 324–328 (2013)
26. HBR. https://hbr.org/2004/04/the-ambidextrous-organization
27. BMC. https://canvanizer.com/new/business-model-canvas
28. Ash, M.: Why Lean Canvas vs Business Model Canvas? https://leanstack.com/why-lean-canvas/
29. IBM. https://www.ibm.com/smarterplanet/us/en/ibmwatson/developercloud/doc/personality-insights/references.shtml#costa1992
30. IBM. https://www.ibm.com/smarterplanet/us/en/ibmwatson/developercloud/doc/personality-insights/basics.shtml
31. IBM. https://www.ibm.com/smarterplanet/us/en/ibmwatson/developercloud/doc/personality-insights/resources/PI-Facet-Characteristics.pdf
32. https://www.ibm.com/smarterplanet/us/en/ibmwatson/developercloud/doc/personality-insights/models.shtml
33. IBM. https://www.ibm.com/cloud-computing/bluemix/watson/
34. Rauhala, L.: Ihminen kulttuurissa – kulttuuri ihmisessä. Gaudeamus (2009)
35. Rauhala, L.: Tajunnan itsepuolustus. Yliopistopaino (1995)

Information Management in the Daily Care Coordination in the Intensive Care Unit

Laura-Maria Peltonen[1($\boxtimes$)], Heljä Lundgrén-Laine[2],
and Sanna Salanterä[1]

[1] Nursing Science, University of Turku, Turku University Hospital,
Turku, Finland
{laura-maria.peltonen, sanna.salantera}@utu.fi
[2] Turku University Hospital, Turku, Finland
helja.lundgren-laine@tyks.fi

Abstract. This study aimed to explore intensive care shift leaders' daily care coordination related information needs and sources of important information. Information needs were explored with a survey and sources of important information were determined through interviews. The survey response rate was 21 % (n = 20) and nine shift leaders were interviewed. The findings are that charge nurses and physicians in charge have different responsibilities and differing information needs, though, some managerial activities and information needs are shared. Shift leaders use numerous sources to obtain necessary information to support decision-making. Sources were categorized into electronic sources, human sources, manual sources and real-time events. Further, information was located within or outside the ICU or based on the shift leader's knowledge. Information systems should be developed based on shift leaders' information needs regardless of the source to support daily care coordination related decision-making. Shift leaders could further benefit from a real-time hospital-wide shared situational awareness.

Keywords: Intensive care · Care coordination · Information needs · Information management · Multidisciplinary · Decision-making

1 Introduction

Intensive care is provided to critically ill patients who suffer from one or more vital organ failure with a life threatening character. Intensive care units (ICUs) are complex and stressful environments where multiprofessional collaboration is vital [1]. This study focuses on decision-making and information management related to the daily care coordination in the ICU.

Managerial decision-making and information management can be described through strategic, tactical and operational levels [2, 3]. The strategic level refers to long-term planning, the tactical level to planning of concrete functions with a shorter-term goal, and the operational level to planning and execution of functions on a daily base. Information management is interconnected to each of these levels horizontally within

© Springer International Publishing Switzerland 2016
H. Li et al. (Eds.): WIS 2016, CCIS 636, pp. 238–252, 2016.
DOI: 10.1007/978-3-319-44672-1_19

and between units, vertically in the organizational hierarchy, and continuously, when specific information is related to the instant, according to Murtola et al. [4].

Care coordination is defined by McDonald et al. [5] as *"the deliberate organization of patient care activities between two or more participants (including the patient) involved in a patient's care to facilitate the appropriate delivery of health care services. Organizing care involves the marshalling of personnel and other resources needed to carry out all required patient care activities, and is often managed by the exchange of information among participants responsible for different aspects of care."* The professionals responsible for the daily care coordination in the ICU are shift leaders, that is, charge nurses and physicians in charge.

A fluent flow of ICU activities requires a large amount of complex combinations of immediate decisions. This care coordination is a complex task as a result of critically ill patients, complex care procedures, constantly changing situations and many professionals involved. The daily care coordination related decisions concern patient admission, specific treatments, material resources, adverse events, human resources, administrative matters, staff knowledge and skills, patient information, medical information, clinical examination findings, diagnoses and patient discharge [6].

1.1 Decision-Making Responsibilities and Information Needs in the Daily Care Coordination

Care coordination in the ICU is conducted in collaboration with the charge nurse and the physician in charge [7, 8]. The charge nurses' responsibilities include to plan and decide up on the allocation of available nursing and material resources based on patient care needs. The physician in charge is responsible for the patients' medical care and in most ICUs, this physician decides about patient admission to the ICU. A difference between ICU shift leaders' decision-making has been reported as the physicians' decision-making related more to patient care when compared to charge nurses' decision-making, which to a great extent was associated with the allocation of resources [8–10].

Information needs related to care coordination concern for example information about the number of incoming patients [8, 9], available material resources [8–10], staff educational background [9, 11], staff experience and critical care skills [9, 11, 12], staffing and skill mix [9, 11, 12], patients and care needs [8–13], and availability of supporting resources [9].

Lundgrén-Laine et al. [9] divided shift leaders' information needs into six dimensions: patient admission, organization and management of work, allocation of staff resources, allocation of material resources, special treatments and patient discharge. They found 57 most crucial information needs for shift leaders and more than one third (22/57) of these were shared between charge nurses and physicians in charge. Furthermore, Lundgrén-Laine et al. [14] compared ICU charge nurses' information needs in Finland and Greece, and found 20 most important information needs that were shared. Most of these (13/20) were staff-related and patient-related connected to the organization and management of work.

1.2 Shift Leaders' Information Management

In user studies, according to Wilson [15], information seeking behavior can be described as starting with an information user who has an information need, which launches an information seeking behavior. The information may be a fact, an advice or an opinion, which may take documented or oral form.

Charge nurses have been found to collect data from several information sources to support their decision-making [8, 16]. Gurses et al. [17] found the existing technology to be insufficient in supporting charge nurses' information needs and hence clinicians assembled paper-based information tools. Further, Lundgrén-Laine et al. [9] found that charge nurses manually updated their information tools several times during a shift. Nonetheless, several information sources are also used by physicians and information systems have shown the potential to reduce the time used for care coordination related information management [18].

Reddy et al. [19] described the importance of the multidisciplinary team members' understanding of the ICU work rhythms as part of a collaborative information seeking behavior to obtain information at the right time when it was needed, not before and not after. Shared information seeking of multidisciplinary ICU teams has also been reported by Gorman and colleagues [20]. Already at that time, they acknowledged a need for digital information management tools to support decision-making.

ICU shift leaders need a vast amount of information and therefore good information management is essential. Furthermore, the format in which information is presented has an effect on both nurses' and physicians' work in the ICU [21]. Effective synthesizing of information can improve communication and decision-making [22] and many information systems have been developed to support decision-making. However, the impact of these systems on managerial decision-making and care processes is yet unclear [4]. The existing body of literature covers to some extent decision-making and information needs in the daily care coordination in the Unites States of America [10, 12], Australia [11, 13], and Europe [7–9, 14]. Research in other countries is needed to verify the generalizability of these findings and to explore possible differences between countries. Yet research exploring ICU shift leaders' sources of information and information management overall is scarce.

1.3 Purpose of the Study

The aim of the study is to explore ICU shift leaders' daily care coordination related information needs and sources of important information. The informants are ICU shift leaders, that is, physicians and nurses in charge, who are responsible for coordinating the daily activities in the ICU during a specific shift. The research questions are:

1. What information do shift leaders need for their immediate decision-making when coordinating care in the ICU and how necessary is this information?
2. What are the sources of the most important information?

2 Methods

2.1 Sample and Setting

This descriptive study was conducted in New Zealand. The study focused on level three ICUs defined by Valentin and Ferdinande [23], to explore information management in the most complex intensive care environments. The ICUs, which were included in the study needed to provide comprehensive care for critically ill adult patients at any time of day, have a designated full-time physician and a senior nurse responsible for the overall management of staff, standards of clinical service provision, and have an academic teaching mission.

All ICUs in New Zealand known by the Intensive Care Nurses Association in New Zealand were contacted and ICUs suiting the inclusion criteria were invited to participate in the study. In total, 29 ICUs were approached from 15 different district health boards. Many Nurse Managers responded to the email but few were from level three ICUs. Five nurse managers from level three ICUs from four district health boards were willing to participate. Two level three ICUs were not interested in participation. The sampling technique was purposive, as the informants needed to have experience of being in charge of daily care coordination in the ICU.

2.2 Data Collection and Analysis

Data were collected in two phases. First, an online survey was used to answer research question one, and second, online interviews were conducted to answer research question two. The interest in phase one lied on the operational decision-making and the order of importance of shift leader's information needs when coordinating care in the ICU. We used a validated questionnaire developed for this purpose by Lundgrén-Laine in 2009. The sources of important information were studied in phase two when the order of importance of the shift leaders' information needs was known.

Data were collected in the first phase with an online questionnaire (Webropol®) between January and May in 2012. The questionnaire was validated in a study by Lundgrén-Laine et al. [9]. It contained seven demographic questions and 122 statements concerning information needed in the decision-making when coordinating care. The scale ranged from 0 (completely unnecessary) to 10 (absolutely necessary). The survey was sent to 95 ICU shift leaders, including charge nurses (N = 61) and physicians in charge (N = 34). The results were used as basis for the interviews in phase two.

Medians were calculated to determine the most important information needs of the respondents. The response rate to the survey was 21 % (n = 20). The medians calculated, provided an order of importance of the information needs. The number of participating physicians remained small (n = 5), and therefore, a larger dispersion between the physicians' important information needs existed in comparison to the charge nurses' information needs. The most important information needs were considered to have a median of 9 and 10 for the charge nurses and a median of 10 for the physicians in charge. This was to ensure that almost all participants considered the information to be absolutely necessary.

Phase two strived to resolve from which sources the most important information is obtained. The data were collected through interviews, which included questions of background information about the shift leader, the unit, and the responsibility and accountability activities of the shift leaders on the unit. The rest of the interview had a structure derived from the results of the questionnaire and was different for charge nurses and physicians in charge as it was based on their specific information needs.

The interview questions for both groups of shift leaders were determined and pilot tested with six shift leaders, including one physician in charge and five charge nurses from one ICU in Finland. Also, two Webropol® questionnaires were built from interview questions for participants unable or unwilling to participate in an online 'live' interview. After responding to this questionnaire an e-mail exchange occurred between the researcher and the participant containing more detailed information about responses to questions in the questionnaire. These questionnaires were language checked and pilot tested with one charge nurse and one physician in charge. Both pilot tests led to minor changes related to the questionnaire setup and additional instructions.

Interview data were collected between June and November in 2012. Nurse Managers or named coordinators of participating units distributed the information letters concerning the study and links to the survey in their unit. In this way the participants were thought to be able to participate without feeling pressured and employees email addresses remained confidential. Participants were asked to provide an email address at the end of the survey if they were willing to take part in an interview. The interviews were estimated to take 30–40 min and they were conducted by the researcher through software allowing communication through voice, video and writing on the Internet such as Skype®, Connect Pro® and email.

Data collected with interviews were analyzed with thematic content analysis. Content analysis is concerned with more than just counting as it considers meanings and context as well [24]. A concept map was used to facilitate understanding of the whole information management process. Graneheim's and Lundman's [25] work guided the content analysis. Data was first read through to get an idea of the whole data set. Then similar expressions were put together into sub categories. After this, upper categories were formed of sources that were similar and finally main categories were defined. Interview data of physicians and nurses were analyzed separately but the results are presented together due to the small sample size.

2.3   Ethical Review and Organizational Permissions

The need for ethical approval was reviewed by the national coordinator of ethics committees at the Ministry of Health in New Zealand based on the Ethical Guidelines for Observational Studies (National Ethics Advisory Committee 2012) in New Zealand. As the nature of the study was observational and posed only minimal risk to participants, no further evaluation by the national ethics committee was needed. Institutional research approval was applied from all organizations participating in the study.

3   Results

The response rate of the survey was 25 % (n = 15) for charge nurses and 15 % (n = 5) for physicians in charge. All ICUs were mixed units, treating both surgical and medical patients. The mean age of survey participants was 47 years. Six informants were men and fourteen were women. Their mean working experience in intensive care was 18 years, with a variation from 7 to 30 years. Seventeen of the informants worked as a shift leader more than three times a week. One informant worked as shift leader on average once a week and two worked two to three times a month. Nine shift leaders, including charge nurses (n = 7) and physicians in charge (n = 2), agreed to an interview. Shift leaders were interviewed from all five ICUs that participated in the study.

3.1 Activities of Shift Leaders

The shift leaders were responsible for ensuring safe and fluent care processes during their shift. Charge nurses and physicians in charge had different responsibilities but some managerial activities were similar, such as supporting staff and allocating work. Charge nurses reported more activities related to the organization and management of work when compared to the physicians in charge who reported more activities related to the patients' care. The shift leaders' responsibilities mainly concerned short-term decision-making and daily care coordination, however, some decisions influenced a longer-term time period continuing beyond the ongoing shift, such as quality assessment of delivered care and professional development. The shift leaders worked closely together to ensure good communication, to improve exchange of information and to ensure a fluent flow of care processes and patient transfers. Examples of the daily care coordination related responsibilities of shift leaders in the ICU are described in Table 1.

Table 1. Examples of activities of shift leaders in the daily care coordination in the ICU.

Charge nurses	Physicians in charge
• Maintaining daily rosters, allocating patients to nurses, and arranging nurse staffing	• Allocating tasks to doctors
• Updating information, such as the on call list, the allocation list, the development requirements, the clinical skills list, and keeping the patient acuity electronic database up to date to reflect true staffing needs	• Deciding on responsible persons for special services
• Coordinating patient movement through the unit with the operating theatre, the emergency department and the wards	• Mentoring and supervising junior doctors
• Contacting wards and units for the transfer of patients	• Managing resources and clarifying available nursing recourses with the charge nurse
• Contacting allied health teams and other services	• Overseeing the daily management of all the patients on the unit
• Supporting staff nurses	• Triaging patients on the phone
• Monitoring care delivery	• Assessing referred patients
• Ensuring appropriate patient isolations	• Deciding what to do if there are insufficient resources to care for all patients in need of intensive care
• Managing maintenance and equipment	• Following up the patients that have been discharged during the last few days
• Dealing with adverse events and complaints	• Liaising with other hospitals about patients, deciding patient transportations and coordinating transportations within and outside of the hospital
	• Communicating with patient's families'

Charge nurses key tasks included provision of professional clinical leadership, ensuring delivery of high quality patient care, demonstrating professional nursing leadership and promoting nursing. They were also responsible for the professional development on the unit and accountable for staff. An important part of their work was to communicate with all nurses as well as with the whole multidisciplinary team in the unit and beyond.

The main responsibilities and accountability activities of the physicians in charge concerned the patients' medical care. However, activities related to the organization and management of work were also reported. This included for example prioritizing and planning for care of the day and bed space use, reporting through medical rounds and handoffs, communication with other professionals and units, and ensuring appropriate admissions and discharges. They were further responsible for appropriate documentation and checking the appropriateness of documentation on the patient's 24-h chart and medication chart. Additionally, they ensured that all patients had been reviewed at least once a day and that the appropriate things and treatments had been conducted. They could furthermore have other responsibilities, which were not directly related to the daily care coordination, such as teaching through bedside or formal sessions with registrars and nurses, and attending senior doctors' meetings for things such as quality assessment and auditing staff.

3.2 Information Needs of Shift Leaders

We identified 42 most important information needs of charge nurses and 39 information needs of physicians in charge. Altogether, 60 % (n = 25) of these were shared between the shift leaders. The distribution between the responses concerning the most important information was small even though the number of respondents remained small. There was a difference between the most important information needed by charge nurses and physicians in charge. Both had most important information needs concerning patient admission, organization and management of work, allocation of staff resources, special treatments and discharge. In addition, charge nurses most important information needs included information about allocation of material resources. The shift leaders most important information needs are presented in Table 2.

3.3 Information Sources of Shift Leaders

Seven charge nurses and two physicians in charge were interviewed concerning the sources of the most important information needs. These interviews included shift leaders from all five ICUs. The collected interview data contained about 50 pages (A4) of transcribed text, with an additional estimated number of 80 pages of email communication and email attachments.

Our thematic content analysis of the information sources resulted in four main categories. These were electronic sources, human sources, manual sources and real time events. These sources were located within or outside of the ICU or were based on the shift leaders experience and knowledge. Both charge nurses and physicians in charge used these sources. The information sources are illustrated in Fig. 1.

Table 2. The most important information needs of shift leaders in the daily care coordination.

Dimension of decision-making	Information need	Charge nurses	Physicians in charge
When a new patient is admitted to the ICU	The number of planned patients	x	x
	The procedures for planned patients	x	x
	The patient's diagnosis		x
	The need to isolate a patient	x	x
	The method of patient isolation	x	x
	The urgency of a patient's condition	x	x
	The patient's name		x
	The patient's earlier physical capacity		x
	The patient's personal identity code		x
	The patient's need for mechanical ventilation	x	x
	The criteria for patient's admission		x
	The emergency operations	x	
Organization and management of work	Staff absence due to sickness	x	
	New patients admitted to ICU	x	x
	New patients presented for admission to ICU	x	x
	The reasons why a patient was refused admission to ICU		x
	Removal of a patient from isolation		x
	Workloads at the unit	x	
	Compulsory infection samples	x	x
	Special treatments given to patients	x	x
	Patient medications that require intensive monitoring	x	x
	A significant change in the patient's condition during one's shift	x	x
	Number of patients on the unit	x	x
	Scheduled examinations that will require patient transfer	x	
	Complications arising during intensive care	x	x
	Adjustments made to equipment supporting a patient's vital functions	x	x
	A patient's death	x	x
	Staff skills and knowledge	x	
	Nursing staff skill mixes	x	
	A patient's allergy	x	x
	Dosages of medications that require intensive monitoring	x	x
	Infusion rate for patient medications that require intensive monitoring		x

(Continued)

Table 2. (*Continued*)

Dimension of decision-making	Information need	Charge nurses	Physicians in charge
	Changes in patient medications that require intensive monitoring		x
	Start time of patient hydration that requires intensive monitoring		x
	Imaging results		x
	Complications related to the patient's diagnosis	x	x
	A patient's intensive care diagnosis/diagnoses	x	x
	Procedures in case of adverse events		x
	Nurse in charge		x
	Physician in charge		
Allocation of staff resources	Number of nursing staff per patient	x	
	Staff resources that can be released	x	
	Staffing for scheduled rosters	x	
	Skill mix on current shift	x	x
	Staffing level on current shift	x	x
	Staff roster changes	x	
	Backup staff	x	
Special treatments	Special treatments	x	x
	Scheduled dates for surgery or procedures	x	
Allocation of material resources	Products needed for special treatments	x	
	Vacant beds on the unit	x	
When a patient is discharged from the ICU	Transport cancellations	x	x
	A patient being discharged	x	
	A patient's time of discharge	x	
	The planned time of transport	x	x
	The family members have been notified of transfer	x	

Shift leaders combined information from several sources to support their decision-making in the daily care coordination. Some information only existed as knowledge in memory or was based on experience. Charge nurses could assemble important information on their own report sheet, which was constructed based on the ICU bed spaces. This sheet had room for notes for all patients including details concerning vital organ function and need for vital organ support. It further included information concerning staffing, patient flow and important phone numbers. This sheet was manually updated several times during the shift.

Information management problems were caused by wrongly documented information, inaccurate information, missing information, a vast amount of informal information, forgotten information, misplaced information and difficulty with information

Table 3. The sources of shift leaders' important information in the daily care coordination.

Main information source	Information system	Examples
Electronic sources	Care related information systems	Allergy alert system, early warning score system, electronic patient management system, infection alert system, patient database, patient information systems, picture archiving and communication system
	Communication systems	Alert system, electronic event form, email, email bookings, internet, intranet, personal computer, senior nurses drive
	Resource allocation systems	Bed management system, computer program
	Staff allocation systems	Nursing workload tool, patient acuity system, payroll system, roster, workload measurement tool
Human sources	Individuals	Family, nurse (e.g. bedside nurse, charge nurse, ward nurse), other health professional (e.g. pharmacist, receptionist), patient, physician (e.g. physician in charge, surgeon)
	Teams	Bed meeting, infection prevention and control team, medical team, senior nurses meeting, patient at risk service, referring team, retrieval team
Manual sources	Communication boards	Pin board at bed space, unit pin board, whiteboard, charge nurses' book
	Management books	Admissions book, clinical allocations book, diary, roster, record of mandatory competencies, calendar for education, staff request book, staffing book, staffing sheet
	Manuals	Charge nurses' manual, clinical pathway, drug manual, guidelines, handover folder, incidents manual, standard policies
	Notes	Admission notes, charge nurse's notes, charge nurses reporting sheet, competency list, staff skills list, own notes, patient list, sheet of paper, staffing sheet
	Patient documentation	Patient chart, medical records, medical prescriptions, medical alert bracelets, treatment prescription chart, care plan
Real time events	A sudden observation	Alarm sign, assistance bell, examination of patient, intercom, observing care equipment, visual sighting
	Planned rounds	Charge nurse round, coordinator round, medical round, routine screening, ward round

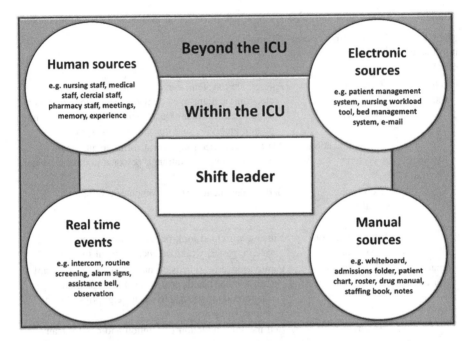

Fig. 1. Information sources of ICU shift leaders in the daily care coordination.

technologies. A need for more accurate information and improvements in information management was recognized. The sources of important information are presented in Table 3.

4 Discussion

The aim of the study was to explore ICU shift leaders' daily care coordination related information needs and sources of important information. The findings are the following. Charge nurses and physicians in charge reported different responsibilities but some managerial activities were similar between charge nurses and physicians in charge. These included allocating work, supporting staff, ensuring quality of care, liaising with others and managing patient flow. The differences were related to the management of material resources, which seemed to only concern the charge nurses, whilst patient medical care related decisions only concerned the physicians in charge. The shift leaders' responsibilities mainly concerned short-term decision-making in care coordination, however, some decisions influenced a longer time period continuing beyond the ongoing shift. This is a new aspect when compared to similar research conducted in Europe [7] and in the United States of America (USA) [10]. Further, charge nurses and physicians in charge shared a great deal (60 %) of important information and they used a variety of information sources to obtain this information to support their decision-making. Used information sources were electronic sources, human sources, manual sources and real time events. The sources were located within or outside of the ICU or were based on the shift leaders experience and knowledge.

The charge nurses' important information needs concerned all six decision-making dimensions whilst physicians in charge had information needs concerning five dimensions, all except the allocation of material resources. This is different when compared to research conducted in Finland [9], where the physicians' information needs only concerned three out of six dimensions, and Greece [14] where charge nurses' information needs concerned five out of six dimensions. In our study, most of the shift leaders' important information needs concerned the organization and management of work. Similar findings were reported in Finland and Greece for charge nurses but not for physicians in charge [9, 14]. The differences in our findings of physicians' information needs may be explained by the small sample size. However, it is clear that charge nurses and physicians in charge have differing information needs, therefore, they would benefit from flexible information management tools that are developed based on their specific information needs.

On the other hand, ICU shift leaders' activities and specific information needs seem alike in different countries. When compared to research conducted in Europe, charge nurses in New Zealand shared 16 out of the 20 most important information needs with charge nurses from Finland and Greece [14]. Further, charge nurses and physicians in charge in New Zealand shared 41 out of the 57 most important information needs with shift leaders from Finland [9].

The most necessary information needed by the ICU shift leaders in daily care coordination was obtained from several sources in different locations. These findings are in line with the work of Wilson [15] concerning user studies. However, in our study the shift leaders additionally obtained important information from real-time events. This new finding should be acknowledged when developing shift leaders' information management. Further, as the ICU shift leaders' information needs and information sources reach beyond their own unit, it seems logic to have a shared situational awareness in the whole care process of the critically ill and not only within the ICU.

Shift leaders reported difficulties with obtaining needed information. Problems with information management might be explained by information systems that do not support shift leaders' information management in daily care coordination related decision-making. Therefore, shift leaders use for example different manual reporting sheets, which they update several times a day with information collected from various sources. Similar findings have been reported both in the USA [17] and in Europe [9]. Generating and updating manual information tools is time consuming [17] and the accuracy of information can be jeopardized as for example information transcription errors may occur. Furthermore, the impact of these tools on care coordination related decision-making is difficult to assess later on as these tools are disposed after use.

Based on the findings of this study, shift leaders' information management is multifaceted and seems to be similar in developed countries in different parts of the world. To date, shift leaders struggle with obtaining necessary information to support their daily care coordination related decision-making. Their need for improved information management is apparent. Shift leaders would benefit from real-time information concerning their most important information needs with easy access and an advanced visualization.

4.1 Limitations

The biggest weakness of the study is the small number of participants. Hence, the results are not generalizable. This study still describes the information sources of the most important information, and even though the results are not conclusive, they provide information of seven intensive care shift leaders' sources of most important information from five different ICUs in different geographical locations in New Zealand on the north and south islands. The participating units also varied in size from small to large and the participants' characteristics varied concerning age, sex, years of working experience and number of days per month when working as shift leader. Therefore, the participants represented a wide range of shift leaders. The internal validity of the questionnaire has been good in a previous study [9] and the interview questions were pilot tested beforehand. Also the questionnaire was pilot tested and language checked before use.

4.2 Significance of Research

The results of this study can be used in further research to improve information management related to the daily care coordination in ICUs. This study provides novel information about the sources of the most important information of shift leaders to be used in the development of information systems to support shift leaders' information management and improve care coordination in the ICU. The study also provides confirmation about similarities of daily care coordination related information management in the ICU in different countries across the globe.

In the future, a larger sample size is needed to validate the preliminary findings of this study. Furthermore, research should be extended to explore shift leaders' information management in the whole critical care setting and beyond, because the development of information and communication technology has the potential to improve care coordination.

5 Conclusions

Charge nurses and physicians in charge have different decision-making responsibilities when coordinating daily care in the ICU and their information needs differ. However, joint managerial activities and information needs also exist, and it seems that at least charge nurses' information needs are alike in different countries. As charge nurses and physicians in charge have differing information needs they can benefit from flexible information management tools developed based on these specific needs. Information systems that support care coordination related decision-making with real-time information are needed and these should be flexible enough to fulfil multidisciplinary demands.

Shift leaders use diverse information sources, which can be categorized into electronic sources, human sources, manual sources and real-time events. As shift leaders' sources of the most important information are located both within the ICU and beyond they could benefit from a shared situational awareness with a hospital-wide information management approach to the care process of the critically ill. Further, information system development and implementation processes should acknowledge shift leaders' information needs regardless of the source of this information.

Acknowledgements. This study was supported by The Finnish Work Environment Fund [grant number 114249].

References

1. Moreno, R.P., Singer, B., Rhodes, A.: What is an ICU? In: Flaatten, H., Moreno, R.P., Putensen, C., Rhodes, A. (eds.) Organisation and Management of Intensive Care, pp. 7–14. Medizinisch Wissenschaftliche Verlagsgesellschaft, Berlin (2010)
2. Winter, A.F., Ammenwerth, E., Bott, O.J., Brigl, B., Buchauer, A., Gräber, S., Grant, A., Häber, A., Hasselbring, W., Haux, R., Heinrich, A., Janssen, H., Kock, I., Penger, O.S., Prokosch, H.U., Terstappen, A., Winter, A.: Strategic information management plans: the basis for systematic information management in hospitals. Int. J. Med. Inform. **64**(2–3), 99–109 (2001)
3. Winter, A., Haux, R., Ammenwerth, E., Brigl, B., Hellrung, N., Jahn, F.: Health Information Systems: Architectures and Strategies, 2nd edn. Springer, New York (2011)
4. Murtola, L.-M., Lundgrén-Laine, H., Salanterä, S.: Information systems in hospitals: a review article from a nursing management perspective. IJNVO **13**(1), 81–100 (2013)
5. McDonald, K.M., Sundaram, V., Bravata, D.M., Lewis, R., Lin, N., Kraft, S., McKinnon, M., Paguntalan, H., Owens, D.K.: Care coordination. In: Shojania, K.G., McDonald, K.M., Wachter, R.M., Owens, D.K. (eds.) Closing the Quality Gap: A Critical Analysis of Quality Improvement Strategies, vol. 7. Technical Review 9 (Prepared by the Stanford University-UCSF Evidence-based Practice Center under contract 290-02-0017). AHRQ Publication No. 04(07)-0051-7. Agency for Healthcare Research and Quality, Rockville (2007)
6. Lundgrén-Laine, H., Salanterä, S.: Think-aloud technique and protocol analysis in clinical decision-making research. QHR **20**(4), 565–575 (2010)
7. Lundgrén-Laine, H., Kontio, E., Perttilä, J., Korvenranta, H., Forsström, J., Salanterä, S.: Managing daily intensive care activities: an observational study concerning ad hoc decision making of charge nurses and intensivists. Nurs. Crit. Care. **15**(4), R188 (2011)
8. Lundgrén-Laine, H., Suominen, H., Kontio, E., Salanterä, S.: Intensive care admission and discharge-critical decision-making points. Stud. Health Technol. Inform. **146**, 358–361 (2009)
9. Lundgrén-Laine, H., Kontio, E., Kauko, T., Korvenranta, H., Forsström, J., Salanterä, S.: National survey focusing on the crucial information needs of intensive care charge nurses and intensivists: same goal, different demands. BMC Med. Inform. Decis. Mak. **13**, 15 (2013)
10. Miller, A., Weinger, M.B., Buerhaus, P., Dietrich, M.S.: Care coordination in intensive care units: communicating across information spaces. Hum. Factors **52**(2), 147–161 (2010)
11. Rischbieth, A.: Matching nurse skill with patient acuity in the intensive care units: a risk management mandate. JNM **14**, 379–404 (2006)
12. Eschiti, V., Hamilton, P.: Off-peak nurse staffing: critical-care nurses speak. DCCN **30**(1), 62–69 (2011)

13. Miller, A., Scheinkestel, C., Joseph, M.: Coordination and continuity of intensive care unit patient care. Hum. Factors: J. Hum. Factors Ergon. Soc. **51**(3), 354–367 (2009)
14. Lundgrén-Laine, H., Kalafati, M., Kontio, E., Kauko, T., Salanterä, S.: Crucial information needs of ICU charge nurses in Finland and Greece. Nurs. Crit. Care **18**(3), 142–153 (2013)
15. Wilson, T.D.: On user studies and information needs. J. Doc. **37**(1), 3–15 (1981)
16. Connelly, L.M., Yoder, L.H., Miner-Williams, D.: A qualitative study of charge nurse competencies. Medsurg. Nurs. **12**(5), 298–306 (2003)
17. Gurses, A.P., Xiao, Y., Hu, P.: User designed information tools to support communication and care coordination in a trauma hospital. J. Biomed. Inform. **42**, 667–677 (2009)
18. Carayon, P., Wetterneck, T.B., Alyousef, B., Brown, R.L., Cartmill, R.S., McGuire, K., Hoonakker, P.L., Slagle, J., Van Roy, K.S., Walker, J.M., Weinger, M.B., Xie, A., Wood, K.E.: Impact of electronic health record technology on the work and workflow of physicians in the intensive care unit. Int. J. Med. Inform. **84**(8), 578–594 (2015)
19. Reddy, M., Pratt, W., Dourish, P., Shabot, M.: Asking questions: information needs in a surgical intensive care unit. In: Proceedings of AMIA Fall Symposium (AMIA 2002), San Antonio, TX, pp. 651–655 (2002)
20. Gorman, P., Ash, J., Lavelle, M., Lyman, J., Delcambre, L., Maier, D., Weaver, M., Bowers, S.: Bundles in the wild: managing information to solve problems and maintain situation awareness. Libr. Trends **49**(2), 266–289 (2000)
21. Miller, A., Scheinkestel, C., Steele, C.: The effects of clinical information on physicians and nurses' decision–making in ICU's. Appl. Ergon. **40**, 753–761 (2009)
22. Feblowitz, J.C., Wright, A., Singh, H., Samal, L., Sittig, D.F.: Summarisation of clinical information: a conceptual model. J. Biomed. Inform. **44**(4), 688–699 (2011)
23. Valentin, A., Ferdinande, P.: ESICM working group on quality improvement. recommendations on basic requirements for intensive care units: structural and organizational aspects. Intensive Care Med. **37**(10), 1575–1587 (2011)
24. Burns, N., Grove, S.K.: The Practice of Nursing Research: Appraisal, Synthesis and Generation of Evidence, 6th edn. Saunders Elsevier, St. Louis (2009)
25. Graneheim, U.H., Lundman, B.: Qualitative content analysis in nursing research: concepts, procedures and measures to achieve trustworthiness. Nurse Educ. Today **24**(2), 105–112 (2004)

Life Satisfaction and Sense of Coherence of Breast Cancer Survivors Compared to Women with Mental Depression, Arterial Hypertension and Healthy Controls

Minna Salakari[1(✉)], Sakari Suominen[1,2], Raija Nurminen[3],
Lauri Sillanmäki[4], Liisa Pylkkänen[5], Päivi Rautava[6],
and Markku Koskenvuo[4]

[1] Department of Public Health, University of Turku, Turku, Finland
minna.salakari@cancer.fi, sakari.suominen@utu.fi
[2] University of Skövde, Skövde, Sweden
[3] University of Applied Science Turku, Turku, Finland
raija.nurminen@turkuamk.fi
[4] Department of Occupational Health, University of Helsinki, Helsinki, Finland
{lauri.sillanmaki,markku.koskenvuo}@helsinki.fi
[5] Cancer Society of Finland, Helsinki, Finland
liisa.pylkkanen@cancer.fi
[6] Turku University Hospital, Turku, Finland
paivi.rautava@utu.fi

Abstract. The purpose of the study was to compare the life satisfaction (LS) and sense of coherence (SOC) of women recovering from breast cancer (BC) to LS and SOC of women with depression or hypertension and of healthy controls. Finnish Health and Social Support (HeSSup) follow-up survey data in 2003 was linked with national health registries. BC patients were followed up for mortality until the end of 2012. The statistical computations were carried out with SAS®. There were no significant differences in LS and SOC between the groups with BC, arterial hypertension or healthy controls. Women recovering from BC are as satisfied with their life as healthy controls, and their perceived LS is better and SOC is stronger compared to women with depression. SOC correlated positively ($r^2 = 0.36$, $p < 0.001$) with LS. However, more studies on determinants of the LS are needed for designing and organizing health care services for BC survivors.

Keywords: Breast cancer · Breast cancer survivors · Comparative study · Life satisfaction · Sense of coherence · Quality of life

1 Introduction

Even worldwide, breast cancer (BC) is the second most common cancer and the most frequent cancer among women [1]. In Finland, in accordance with Western statistics [2], over the period 2007–2011, the annual BC incidence was more than 4,500, and the annual incidence rate was 92.6, and the mortality rate 13.8 per 100,000 inhabitants [2].

© Springer International Publishing Switzerland 2016
H. Li et al. (Eds.): WIS 2016, CCIS 636, pp. 253–265, 2016.
DOI: 10.1007/978-3-319-44672-1_20

Life satisfaction (LS) represents a cognitively pronounced assessment of a person's situation in life and how (s)he feels about the future. It is a measure of well-being, and the degree to which a person positively evaluates the overall quality of life (QoL). QoL is a multidimensional concept involving aspects of individuals' overall wellbeing [3], which is a state where essential needs are met and welfare increases in proportion to how far the needs are satisfied [4]. QoL may be assessed for instance in terms of mood, satisfaction with relationships, self-concepts, and self-perceived ability to cope with daily life [3–5]. It constitutes a part of the broader concepts of well-being and overall quality of life (QoL). QoL is a multidimensional concept involving aspects of individuals' overall wellbeing [3]. It may be assessed for instance in terms of mood, satisfaction with relationships, self-concepts, and self-perceived ability to cope with daily life [3–5].

For patients with cancer, LS and QoL has been determined as the patient's appraisal of satisfaction with their current level of functioning as compared with what they perceive to be possible or ideal [6]. The determinants of LS and QoL in women with BC include psychosocial factors, such as coping style, as well as sociodemographic and medical factors [7]. Different coping strategies have distinct effects on LS, QoL and general wellbeing in women with BC [8, 9].

Cancer is one of the most feared diseases in modern societies and shows a strong negative impact on LS and QoL. The diagnosis of cancer evokes the feeling of stress, fear, sorrow, anger and uncertainty [10–12] and clinically significant depression is typically associated with BC [13, 14]. Depression has also been shown to impair general QoL [15] and individuals continue to report high levels of psychological stress and depression, even once cancer treatments are over [16].

Person's sense of coherence (SOC) refers to an individual's orientation towards life and is a central concept of Antonovsky's theory of salutogenesis [17]. SOC may be described as a readiness of an individual to experience life as understandable, manageable and meaningful and to deal with stressors appropriately. Antonovsky's initial research focused on health promoting factors instead of health risks. Antonovsky hypothesized health as a continuum between total disease (dis-ease) and complete health (ease). The concept of salutogenesis challenges the traditional medical view on pathogenesis [17, 18]. SOC associates strongly with perceived LS and QoL [19].

BC patients with a strong SOC experience fewer stressful events and better overall QoL. The associations between, on the one hand, SOC and, on the other hand, health status, LS and QoL are linear [20, 21]. SOC, self-esteem, coping strategies, social support, global meaning and emotional distress are associated with QoL among patients with cancer. There is a positive link between self-esteem and QoL, and fighting spirit impacts positively on QoL [22].

Although the incidence of BC is increasing, the number of younger BC patients and consequently survivors is low and only few studies have addressed the LS and SOC among young BC survivors. Further, no comparative studies on LS between BC patients and other groups of patients suffering from common chronic diseases have been performed.

The purpose of this study was to compare the LS and SOC of women with BC after initial recovery to LS and SOC of two groups of women suffering from common long-term conditions, i.e. mental depression and arterial hypertension, and to LS and SOC of healthy controls. According to the study hypothesis, recovery from BC could be detected as impaired LS and SOC among BC survivors in comparison with healthy controls, where we expected the greatest differences to appear, but we also expected differences in comparison to corresponding female patients with arterial hypertension or mental depression.

2 Methods

2.1 Study Design and Participants

The data of the study is drawn from The Health and Social Support (HeSSup) study which is an ongoing prospective cohort study on a nationwide representative sample of the Finnish population [23]. The study started in 1998 and four initial age groups have been followed up: 20–24, 30–34, 40–44 and 50–54 years. The total number of all participants is 25,895, and female participants 15,267.

The HeSSup survey data of year 2003 was used and, with the consent of the respondents, linked with national registry data from the years 1998–2012 of the Finnish Cancer Registry, the Finnish Drug Purchase and Imbursement Registry of the Social Insurance Institution, and mortality statistics from Statistics Finland. BC patients were followed-up for mortality until the end of 2012. The respondents who had a registered (verification by the Finnish Cancer Registry) diagnosis of BC in 1998–2002, were alive at least one year after diagnosis, and had completed the 2003 survey, formed the study group (N = 56). No woman had to be excluded due to death. Also those women with BC, who also had reported depression, were included. The patients with BC were followed-up for mortality until the end of 2012 when nine respondents with BC had died due to it. There were three study groups to be compared: (1) the respondents who reported having suffered from mental depression in the 2003 survey, had purchased antidepressants and were alive in 2003–2005 (N = 471); (2) the respondents who reported having arterial hypertension in the 2003 survey, had purchased anti-hypertensive medication and were alive in 2003–2005 (N = 841); and (3) all respondents of corresponding age who had not reported any chronic disease, any cancer, depression or hypertension (N = 6274) in the 2003 survey.

Purchase of medication as well as survival was determined according to registry data. The survey data was with consent of the respondents linked with national registry data from the years 1998–2012 of the Finnish Cancer Registry, the Finnish Drug Purchase and Imbursement Registry of the Social Insurance Institution, and mortality statistics from Statistics Finland. Control disease groups with high incidence were selected for the study control group. These diseases are common public health problems among the study population in Finland. The advantage of using several control groups provides a basis for valid generalization. The study design and participants is presented in Fig. 1.

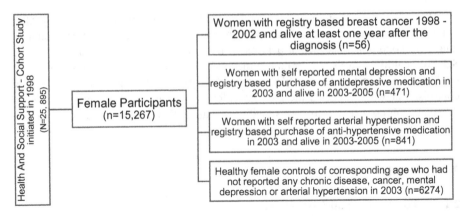

Fig. 1. Study design and participants

2.2 Ethics

Since the HeSSup study was not a medical study, the concurrent joint Ethics Committee of the University of Turku and the Turku University Central Hospital considered that formal approval was not needed and stated that the study followed the ethical guidelines for good scientific practice.

2.3 Outcome Variables

The data on the outcome variables were derived from the 2003 survey. Life satisfaction (LS) is an essential part of QoL. LS was assessed using a four-question scale modified from a questionnaire developed for measuring the QoL in Nordic QoL studies [24]. The participants were asked to rate four aspects of their life satisfaction: interest in life, happiness, ease of living and loneliness (Very interesting/Happy/Easy/Not at all = 1, Very boring/Unhappy/Hard/Lonely = 5). The LS sum score (range: 4–20) was further classified into the following categories: satisfied (4–6), intermediately satisfied (7–11) and unsatisfied (12–20). The sum scale of LS variable was reversed for statistical processing so that a higher score corresponded to better life satisfaction and vice versa. A sum was calculated under the condition that at least three items were responded to. A missing value was replaced by the mean of the remaining three.

SOC was assessed using Antonovsky's 13-item scale [18] which is derived from the original 29-item Orientation to Life Questionnaire. It covers the three main sub-components of SOC: comprehensibility (5 items), manageability (4 items), and meaningfulness (4 items). All items have a seven graded (Likert-type) response scale, and a sum was calculated (range 13–91) under the condition that at least 3 comprehensibility, 2 manageability and 2 meaningfulness items were filled in. Missing values were replaced with the mean of other items in each sub-component.

The respondent's level of education was classified into four categories: no professional education; vocational course or school/apprentice contract: college, and university/other high level education.

2.4 Statistical Analyses

The age group 20–24 years was omitted from the statistical analysis since there was only one respondent with BC.

Overall associations between the variables were measured with the Cochran-Mantel-Haenszel test. The following descriptive statistics were calculated: frequency, mean with 95 % confidence interval, range and median. The ANOVA with pairwise comparison with Tukey-Kramer adjustment was used to determine differences between study groups. Pearson's correlation test was used to compare degrees of association between LS and SOC scale.

The limit for statistical significance was set at $p = 0.05$. Statistical analyses were performed with the SAS® software v.9.3 for Windows (SAS Institute Inc., USA).

3 Results

Respondents' age and educational background is presented in Tables 1 and 2. Most respondents were 40–44 years old, and most of them had vocational course/school or college education (Tables 1 and 2). The number of BC survivors after initial recovery was relatively low, as the age of participants was generally low. The age and educational distributions were consistent in each group which enables the comparisons between different groups.

Table 1. Number and % of row sum of the women studied by age.

Age	30–34 years		40–44 years		50–54 years		All	
	Freq.	%	Freq.	%	Freq.	%	Freq.	%
Breast cancer	3	5.4	20	35.7	33	58.9	56	100
Depression	138	29.3	187	39.7	146	31.0	471	100
Arterial hypertension	70	8.3	215	25.6	556	66.1	841	100
Controls	2330	37.1	2181	34.8	1763	28.1	6274	100
Total	**2541**	**33.3**	**2603**	**34.0**	**2498**	**32.7**	**7642**	**100.0**

3.1 Life Satisfaction (LS)

LS among women with BC, mental depression or arterial hypertension, and among healthy female controls is presented by age group and by age groups combined in Table 3.

When all age groups were combined there were no significant differences in LS between patients with BC, patients with hypertension or healthy participants, while the group with depression had significantly poorer LS compared to the rest (Table 3). All results for LS remained unchanged when adjusted for age and level of education. When analyzed according to age group, the result was particularly apparent among the oldest respondents (Table 3). The statistically significant differences in comparison with the group suffering from depression persisted except for the youngest age group.

Table 2. Number and % of row sum of the women studied by education.

Level of education	No. professional education		Vocational course or school/apprentice contract		College		University or higher		All	
	Freq.	%	Freq.	%	Freq.	%	Freq.	%	Freq.	%
Breast cancer	12	18.8	13	23.2	16	28.6	15	26.8	56	100
Depression	67	14.2	154	32.7	175	37.2	72	15.3	468	100
Arterial hypertension	158	18.8	289	34.4	308	36.6	79	9.4	834	100
Controls	764	12.2	1886	30.1	2519	40.4	1057	24	6226	100
Total	1001	13.2	2344	30.9	3021	39.8	1226	16.1	7582	100

Table 3. Life satisfaction sum scores of women with registry based breast cancer, self-reported and medicated mental depression and self-reported and medicated arterial hypertension and of healthy women controls by age groups and age groups combined. The values in the columns additionally to age and N stand for Range, Mean with 95 % confidence interval (CL), median and observed statistical significance.

Group	Age	N	Range	Mean (95 % CI)	Median
Breast cancer	30–34	3	13.0–14.0	13.7 (12.2–15.1)	14.0
	40–44	20	7.0–19.0	15.2 (13.8–16.5)	16.5
	50–54	33	7.0–20.0	15.8 (14.8–16.9)	17.0
Mental depression	30–34	138	5.0–20.0	13.8 (13.2–14.5)	15.0
	40–44	187	4.0–20.0	13.1 (12.5–13.7)	14.0
	50–54	146	6.0–20.0	13.8 (13.2–14.4)	15.0
Arterial hypertension	30–34	70	8.0–20.0	15.5 (14.8–16.2)	17.0
	40–44	215	5.0–20.0	15.0 (14.6–15.5)	17.0
	50–54	556	4.0–20.0	15.5 (15.2–15.8)	17.0
Controls	30–34	2300	4.0–20.0	15.6 (15.4–15.7)	17.0
	40–44	2181	4.0–20.0	15.5 (15.4–15.6)	17.0
	50–54	1763	4.0–20.0	15.7 (15.6–15.9)	17.0

Age groups combined	N	Range	Mean (95 % CI)	Median	p*
Breast cancer	56	7.0–20.0	15.5 (14.7–16.3)	17.0	Ref.
Mental depression	471	4.0–20.0	13.5 (13.2–13.9)	15.0	0.0001
Arterial hypertension	841	4.0–20.0	15.4 (15.2–15.6)	17.0	0.9971
Controls	6274	4.0–20.0	15.6 (15.5–15.7)	17.0	0.9917

*Tukey-Kramer test for pairwise mean differences between BC and other groups.

3.2 Sense of Coherence (SOC)

Individuals with mental depression had the lowest SOC scores in each age group; the nadir was for the age group 30–34 years. In the younger age groups the differences in SOC between BC survivors and healthy controls were the greatest. The younger age groups had overall lower SOC scores than the older ones. The differences were, however, not statistically significant. The individuals with BC in age group 50–54 years had higher SOC scores than individuals of same age in any other group. The individuals with arterial hypertension had lower SOC scores than individuals in BC group when the age groups were analyzed together. These differences were not statistically significant, except for the depression group (Table 4).

Table 4. Sense of coherence scores of women with registry based breast cancer, self-reported and medicated mental depression and self-reported and medicated arterial hypertension and of healthy female controls by age groups and age groups combined. The values in the columns additionally to age and N stand for Range, Mean, Median, and 95 % confidence interval. CI = confidence interval.

Group	Age	N	Range	Mean (95 % CI)	Median
Breast cancer	*30–34*	*3*	*60.0–62.0*	*61.0 (58.5–63.5)*	*61.0*
	40–44	20	29.0–82.0	62.9 (56.8–69.0)	65.0
	50–54	*33*	*44.0–86.0*	*69.7 (66.1–73.3)*	*69.0*
Mental depression	30–34	138	24.0–86.0	58.3 (56.1–60.4)	59.0
	40–44	*187*	*19.0–91.0*	*58.4 (56.6–60.2)*	*59.0*
	50–54	146	32.0–86.0	60.2 (58.2–62.1)	61.0
Arterial hypertension	*30–34*	*70*	*37.0–87.0*	*66.6 (62.8–68.5)*	*68.0*
	40–44	215	25.0–90.0	65.6 (64.0–67.2)	68.0
	50–54	*556*	*25.0–90.0*	*66.4 (65.5–67.3)*	*67.0*
Controls	30–34	2330	25.0–91.0	66.1 (65.7–66.6)	67.0
	40–44	*2181*	*20.0–91.0*	*67.0 (66.6–67.5)*	*68.0*
	50–54	1763	28.0–90.0	68.3 (67.8–68.8)	69.0
Age groups combined	N	Range	Mean (95 % CI)	Median	p*
Breast cancer	*56*	*29.0–86.0*	*66.8 (63.7–69.9)*	*67.7*	*Ref.***
Mental depression	471	19.0–91.0	58.9 (57.8–60.0)	59.0	<0.0001
Arterial hypertension	*841*	*25.0–90.0*	*66.1 (65.4–66.9)*	*67.0*	*0.9705*
Controls	6274	20.0–91.0	67.1 (66.8–67.3)	67.3	0.9982

Tukey-Kramer test for pairwise mean differences between BC and other groups.

Overall, the healthy controls had the highest SOC scores in all other age groups except age group 50–54 years (Table 4), but the differences were not statistically significant compared to the other groups except for the depression group which, as stated above, had the poorest SOC.

There was a positive correlation ($r^2 = 0.36$, p < 0.001) between SOC and LS, i.e., the higher the SOC, the better the LS.

4 Discussion

Contrary to the study hypothesis the results showed that women with BC in their initial recovery phase, at least one year after diagnosis, did not have impaired LS and SOC. Compared to healthy female controls and the group of women with arterial hypertension there were no significant differences between women with BC and the other groups. However, women with depression reported significantly lower LS and SOC scores than the other groups. These results were particularly apparent when all the age groups were combined.

This is an interesting finding, since BC diagnosis has generally been reported to raise fear of death and distress [10, 18]. Moreover, the sample included also those participants with BC, who also had self-reported depression in the 2003 survey. However, one has to keep in mind that the LS improves with the number of post-diagnosis years [25], and women with BC in this study were young and thus highly selected. Moreover, individuals with the poorest prognosis were possibly not alive one year after diagnosis or may due to this or other reasons not have participated in the second data collection of the HeSSup survey at all. This implies that the disease was for many individuals considered treatable and even completely curable which probably has provided new hope and improved their LS and strengthened their SOC. BC treatment has evolved over the years, and at present the survival of BC patients in Finland is excellent with 90 % 5-years survival [2]. This most likely has an impact on the SOC and LS of the patients. Moreover, most BC patients have also already recovered from the acute crisis the disease with great probability has caused to most of them.

The LS of women with BC in this study was relatively good. It seems that women may have a sense of relief after having survived at least a year after the diagnosis. Ashing-Giwa et al. [26] reported similar findings on improved health status and relatively good overall QoL in African-American and Caucasian BC groups. Also Yost et al. [27] and Hsu et al. [28] have reported good LS among BC survivors. This may be due to the fact that the worst phase of the disease and crisis has passed, and the survivors may have received support services during their disease, which may have contributed to the process of surviving and the maintenance and improvement of LS.

To our knowledge, the design of the present study is unique in that it compares LS and SOC of women in initial recovery phase from BC with LS and SOC of two groups of women suffering from common long-term conditions, i.e. mental depression or arterial hypertension, and with LS and SOC of healthy female controls. The 4-item life satisfaction scale is easily administered and well accepted [23]. The 13-item SOC questionnaire, equally easy to administer, has been applied in a number of population studies [19]. Our study groups were mutually exclusive and the participants were entered into one study group only, except for the women with BC in which women suffering from mental depression were not excluded, whereas depressed individuals were excluded from the hypertension and control groups. Due to small numbers of women diagnosed with BC, the exclusion of those BC patients with depression may have led to biased conclusions. However, in order to explore the potential influence of having included women with mental depression matching the study criteria into the group with BC, additional statistical analyses were performed by excluding these individuals (N = 11) yielding unchanged key results (data not shown).

The patients with BC were followed-up for mortality until the end of 2012 when nine respondents with BC had died due to it. This extended follow-up increases the reliability of the results and confirms that results obtained are not due to selection bias (i.e., only patients with the best prognosis have responded).

BC is a serious disease, and one would assume that the diagnosis would certainly affect the patients' LS and SOC [10, 29, 30]. Younger age in women with BC has been shown to be one of the significant risk factors for poor LS and greater psychological morbidity compared to older women [31, 32]. Many BC survivors report weakened LS and younger BC survivors are at risk for impaired LS up to several years after diagnosis [32]. There is also evidence that mental depression is one of the symptoms or a consequence of the cancer diagnosis [33].

Based on our results, it seems that during the survival phase, after the acute crisis, LS might improve. Cancer, as a frightening disease at the time of diagnosis, may weaken LS, but coping with the disease even improves it - serious illness and the survival becomes as a part of survivors' own history. The conclusion is consistent with previous findings showing that coping with a serious disease leads to hope and gratitude, and growth and empowerment [34]. Coping raises patient's fighting spirit against the disease and improves current LS. The way of coping with fears and long-term effects might be predictors for long-term LS after cancer [34, 35].

More studies are still needed to examine factors affecting the LS, which form an important base for the development of cancer rehabilitation and treatment. BC treatment is currently rather well structured and harmonized in Finland so that rehabilitation is usually offered after treatment, usually more than a year from diagnosis. It has been shown, however, that people with cancer need peer, psychosocial and physical support and information about cancer during the disease [36]. Structured measurement of LS enables comparisons between a variety of diseases and treatments, which is needed for decisions within health care, especially when nationwide health policies for cancer patients are organized.

Untreated mental depression clearly impairs the overall QoL. Depression understandably affects LS negatively and most of the loss of LS associated with chronic diseases may be attributed to mental disorders. The patient's perception of the severity of her disease is related to the complexity of the disease and to a number of areas of life that are affected by the symptoms [15]. These observations explain some of the present findings, e.g., that depression was associated with the poorest LS and SOC, even considering the use of antidepressants. Indeed, treatment of depression does improve LS, but depression is still associated with a poor overall QoL even when the patient is on medication [37].

SOC lies within the domain of a person's health resources and, in that, SOC influences LS and overall QoL [19]. There is substantial evidence that SOC plays a central role in a patient's coping with stressors during rehabilitation or recovery and that SOC contributes to mental health and psychosocial functioning [38]. Our results support Antonovsky's theoretical assumptions of SOC [17]. LS was linearly associated with SOC, i.e., the stronger the SOC, the better the LS. Zielińska-Więczkowska et al. [39] have found that a strong SOC and high level of education have significant effects on LS. In our study, however, the educational background had no significant impact on the results.

The results can only be generalized to young women with BC (aged below 55 years) without any comorbidity. Age is inversely associated with LS [40]. The group of BC patients in this study represents a very special group, as majority of BC patients in general population are over 60 years. Young BC patients may have special needs for rehabilitation and psychosocial support, which needs to be investigated further. In Finland, practically all BC patients, in addition to medical treatment, undergo various supplementary interventions. The study lends support to the efficiency of these interventions. Unfortunately, no information about the interventions on the individual level is available. Hence, more studies on determinants of the efficiency of these interventions are needed.

During the recent decades, 5-year survival rates have improved among cancer patients who are otherwise healthy, but not among individuals with severe comorbidities [41]. Overall QoL and outcomes from most functional and symptom scales seem to weaken over time [29]. Apparently patients with cancer and psychological symptoms and depression have a poorer LS [14] and patients with BC have generally a high incidence of depression and anxiety and a reported low LS [31].

At the baseline HeSSup study, the study participation rate was moderate, only 40 %. Nevertheless, an analysis of non-respondents has shown that the sample regarding health parameters is representative for the Finnish population [23]. The probability of not responding was greater among men, older age groups, those with less education, divorced or widowed respondents and respondents on disability pension. It has been shown that respondents tend to be healthier and report a more favorable health behavior than non-respondents [23]. It is also known that women and those of a higher social class tend to participate more actively in health-related survey research than males and persons of lower social class. This may lead to underestimation of the prevalence of health problems among men and persons of lower social strata. Another theory which explains passive or active responses to health issues is SOC. It has been found to have a strong positive impact on health information literacy [42].

The strengths of the study include the use reliable registry data, so we did not need to rely solely on self-reported data. According to concurrent legislation, The Finnish Cancer Registry registers all cancer cases in Finland and all medication is registered by The Finnish Drug Purchase and Imbursement Registry of the Social Insurance Institution, so the allocation of patients within groups is reliable. Other strengths are the extensive source data of more than 15,000 women, and careful follow-up of the individuals in the study. Even though the BC patient group is small, it can be considered reliable and unbiased. As the number of young (under 55 years) BC survivors is generally low, this places this research data in a unique position.

In summary, after at least initial recovery from BC, LS and SOC do not seem to be permanently affected on a group level since the women studied were rather satisfied with their lives and reported a good level of SOC. Women recovering from BC were in fact as satisfied with their life as healthy controls. The SOC and LS also inter-correlated statistically significantly in accordance with earlier findings. However, factors affecting LS and SOC need to be examined and described more in detail in the future for designing and organizing optimal supportive health care and rehabilitation services for BC survivors.

5 Conclusions

After at least initial recovery from BC, LS and SOC do not seem to be permanently affected on a group level since the women studied were rather satisfied with their lives and reported a good level of SOC. Women recovering from BC were in fact as satisfied with their life as healthy controls. The SOC and LS also inter-correlated statistically significantly in accordance with earlier findings.

However, factors affecting LS and SOC need to be examined and described more in detail in the future for achieving even improved supportive health care and rehabilitation services for BC survivors.

References

1. Globocan 2012. Fast Stats. Most frequent cancers: women. http://globocan.iarc.fr/factsheets/populations/factsheet.asp?uno=900#WOMEN. Accessed 05 Jan 2016
2. Finnish Cancer Registry. http://www.cancer.fi/syoparekisteri/en/statistics/cancer-statistics/. Accessed 10 Jan 2016
3. Cella, D.F.: Quality of life: concepts and definition. J. Pain Symptom Manag. 9, 186–192 (1994)
4. Allardt, E.: Hyvinvoinnin ulottuvuudet. WSOY, Porvoo (1976)
5. Veenhoven, R.: The Study of Life Satisfaction. Eötvös University Press, Budapest (1996). ISBN 963 463 081 2, pp. 11–48
6. Cella, D., Cherin, E.: Quality of life during and after treatment. Compr. Ther. 14, 69–75 (1988)
7. Mols, F., Vingerhoets, A.J., Coebergh, J.W., Poll-Franse, L.V.: Quality of life among long-term breast cancer survivors: a systematic review. Eur. J. Cancer 2005(41), 2613–2619 (2005)
8. Garnefski, N., Kraaij, V., Spinhoven, P.: Negative life events, cognitive emotion regulation and emotional problems. Pers. Individ. Differ. 2001(30), 1311–1327 (2001)
9. Lehto, U.S., Ojanen, M., Kellokumpu-Lehtinen, P.: Predictors of quality of life in newly diagnosed melanoma and breast cancer patients. Ann. Oncol. 16, 805–816 (2005)
10. Yusuf, A., Ahmad, Z., Keng, S.L.: Quality of life in Malay and Chinese women newly diagnosed with breast cancer in Kelantan Malaysia. Asian Pac. J. Cancer Prev. 14, 435–440 (2013)
11. Carlson, L.E., Bultz, B.D.: Cancer distress screening: needs, models, and methods. J. Psychosom. Res. 55(5), 403–409 (2003)
12. Stark, D., Kiely, M., Smith, A., Velikova, G., House, A., Selby, P.: Anxiety disorders in cancer individuals: their nature, associations, and relation to quality of life. J. Clin. Oncol. 20 (14), 3137–3148 (2014)
13. So, W.K.W., Marsh, G., Ling, W.M., Leung, F.Y., Lo, J.C.K., Yeung, M., Li, G.: Anxiety, depression and quality of life among Chinese breast cancer individuals during adjuvant therapy. Eur. J. Oncol. Nurs. 14(1), 17–22 (2010)
14. Howard-Anderson, J., Ganz, P.A., Bower, J.E., Stanton, A.L.: Quality of life, fertility concerns, and behavioral health outcomes in younger breast cancer survivors: a systematic review. J. Natl. Cancer Inst. 104(5), 386–405 (2012)
15. Saarni, S.: Health-related quality of life and mental disorders in Finland. Department of Mental Health and Alcohol Research National Public Health Institute, Helsinki and Department of Psychiatry University of Helsinki, Helsinki, Finland (2008)
16. Holzner, B., Kemmler, G., Kopp, M., Moschen, M., Schweigkofler, H., Dünser, M., Margreiter, R., Fleischhacker, W.W., Sperner-Unterweger, B.: Quality of life in breast cancer individuals—not enough attention for long-term survivors? Psychosomatics 42, 117–123 (2001)

17. Schumacher, J., Wilz, G., Gunzelmann, T., Brahler, E.: The Antonovsky sense of coherence scale. Test statistical evaluation of a representative population sample and construction of a briefscale. Psychother. Psychosom. Med. **50**, 472–482 (2000)

18. Antonovsky, A.: Unraveling the Mystery of Health. How People Manage Stress and Stay Well San Francisco. Jossey-Bass, London (1987)

19. Eriksson, M., Lindström, B.: Antonovsky's sense of coherence scale and its relation with quality of life: a systematic review. J. Epidemiol. Commun. Health 61(11), 938– 944 (2007)

20. Sarenmalm, K.E., Browall, M., Persson, L.-O., Fall-Dickson, J., Gaston-Johansson, F.: Relationship of sense of coherence to stressful events, coping strategies, health status, and quality of life in women with breast cancer. Psycho-Oncology (2011). doi:10.1002/pon.2053

21. Gerasimčik-Pulko, V., Pileckaitė-Markovienė, M., Bulotienė, G., Ostapenko, V.: Relationship between sense of coherence and quality of life in early stage breast cancer individuals. Acta Med. Litu. **16**(3–4), 139–144 (2009)

22. Allart, P., Soubeyran, P., Cousson-Gélie, F.: Are psychosocial factors associated with quality of life in individuals with haematological cancer? A critical review of the literature. Psycho-Oncology **22**(2), 241–249 (2013). doi:10.1002/pon.3026

23. Korkeila, K., Suominen, S., Ahvenainen, J., Ojanlatva, A., Rautava, P., Helenius, H., Koskenvuo, M.: Non-response and related factors in a nation-wide health survey. Eur. J. Epidemiol. **17**, 991–999 (2001)

24. Allardt, E.: About dimension of welfare: an explanatory analysis of the comparative Scandinavian survey. University of the Helsinki, Research Group of Comparative Sociology Research reports, no.1 (1973)

25. Cimprich, B., Ronis, D.L., Martinez-Ramos, G.: Age at diagnosis and quality of life in breast cancer survivors. Cancer Pract. **10**, 85–93 (2002)

26. Ashing-Giwa, K., Ganz, P.A., Petersen, L.: Quality of life of African-American and white long term breast carcinoma survivors. Cancer **85**, 418–426 (1999)

27. Yost, K.J., Haan, M.N., Levine, R.A., Gold, E.B.: Comparing SF-36 scores across three groups of women with different health profiles. Qual. Life Res. **14**(5), 1251–1261 (2005)

28. Hsu, T., Ennis, M., Hood, N., Graham, M., Goodwin, P.: Quality of life in long term breast cancer survivors. Am. Soc. Clin. Oncol. **31**, 3540–3548 (2013)

29. Bloom, J., Stewart, S., Oakley-Girvan, I., Banks, I.: Quality of life of younger breast cancer survivors: persistence of problems and sense of well-being. Psycho-Oncology **21**(6), 655–665 (2012)

30. Helgeson, V.S., Snyder, P., Seltman, H.: Psychological and physical adjustment to breast cancer over 4 years: identifying distinct trajectories of change. Health Psychol. **23**(1), 3–15 (2004)

31. Kwan, M.L., Ergas, I.J., Somkin, C.P., Quesenberry Jr., C.P., Neugut, A.I., Hershman, D.L., et al.: Quality of life among women recently diagnosed with invasive breast cancer: the pathways study. Breast Cancer Res. Treat. **123**, 507–524 (2010)

32. Avis, N.E., Crawford, S., Manuel, J.: Quality of life among younger women with breast cancer. J. Clin. Oncol. **23**, 3322–3330 (2005)

33. Weitzner, M.A., Meyers, C.A., Stuebing, K.K., Saleeba, A.K.: Relationship between quality of life and mood in long-term survivors of breast cancer treated with mastectomy. Support. Care Cancer **5**, 241–248 (1997)

34. Chopra, I., Kamal, K.: A systematic review of quality of life instruments in long-term breast cancer survivors. Health Qual. Life Outcomes **10**, 14 (2012). doi:10.1186/1477-7525-10-14

35. Dahl, L., Wittrup, I., Væggemose, U., Petersen, L.K., Blaakaer, J.: Life after gynecologic cancer-a review of individuals quality of life, needs, and preferences in regard to follow-up. Int. J. Gynecol. Cancer **23**(2), 227–234 (2013). doi:10.1097/IGC.0b013e31827f37b0

36. Dilworth, S., Higgins, I., Parker, V., Kelly, B., Turner, J.: Individual and health professional's perceived barriers to the delivery of psychosocial care to adults with cancer: a systematic review. Psycho-Oncology **23**(6), 601–612 (2014). doi:10.1002/pon. 3474

37. IsHak, W.W., Greenberg, J.M., Balayan, K., Kapitanski, N., Jeffrey, J., Fathy, H., Fakhry, H., Rapaport, M.H.: Quality of life: the ultimate outcome measure of interventions in major depressive disorder. Harv. Rev Psychiatry **19**(5), 229–239 (2011)

38. Griffiths, C.A.: Sense of coherence and mental health rehabilitation. Clin. Rehabil. **23**(1), 72–78 (2009). doi:10.1177/0269215508095360

39. Zielińska-Więczkowska, H., Ciemnoczołowski, W., Kędziora-Kornatowska, K., Muszalik, M.: The sense of coherence (SOC) as an important determinant of life satisfaction, based on own research, and exemplified by the students of University of the Third Age (U3A). Arch Gerontol. Geriatr. **54**(1), 238–241 (2012). doi:10.1016/j.archger.2011.03.008

40. Knobf, M.: Psychosocial responses in breast cancer survivors. Semin. Oncol. Nurs. **23**(1), 71–83 (2007)

41. Søgaard, M., Thomsen, R., Bossen, K., Sørensen, H., Nørgaard, M.: The impact of comorbidity on cancer survival: a review. Clin. Epidemiol. **5**(1), 3–29 (2013). doi:10.2147/ CLEP.S47150

42. Ek, S., Widén-Wulff, G.: Information mastering, perceived health and societal status: an empirical study of the Finnish population. Libri **58**(2), 74–81 (2008)

Author Index

Printed in the United States
By Bookmasters